A Level Chemistry for OCR Year 1 and AS

A

Rob Ritchie • Emma Poole
Series Editor: Rob Ritchie

OXFORD
UNIVERSITY PRESS

Great Clarendon Street, Oxford, OX2 6DP, United Kingdom

Oxford University Press is a department of the University of Oxford. It furthers the University's objective of excellence in research, scholarship, and education by publishing worldwide. Oxford is a registered trade mark of Oxford University Press in the UK and in certain other countries

© Rob Ritchie and Emma Poole 2017

The moral rights of the authors have been asserted

First published in 2017

All rights reserved. No part of this publication may be reproduced, stored in a retrieval system, or transmitted, in any form or by any means, without the prior permission in writing of Oxford University Press, or as expressly permitted by law, by licence or under terms agreed with the appropriate reprographics rights organization. Enquiries concerning reproduction outside the scope of the above should be sent to the Rights Department, Oxford University Press, at the address above.

You must not circulate this work in any other form and you must impose this same condition on any acquirer

British Library Cataloguing in Publication Data
Data available

978-0-19-835198-6

10 9 8 7 6 5 4 3 2 1

Paper used in the production of this book is a natural, recyclable product made from wood grown in sustainable forests. The manufacturing process conforms to the environmental regulations of the country of origin.

Cover: SCIENCE PHOTO LIBRARY

Artwork by Q2A Media

Printed and bound in Great Britain by Bell and Bain Ltd, Glasgow

AS/A Level course structure

This book has been written to support students studying for OCR AS Chemistry A and for students in their first year of studying for OCR A Level Chemistry A. It covers the AS modules from the specification, the content of which will also be examined at A Level. The modules covered are shown in the contents list, which also shows you the page numbers for the main topics within each module. If you are studying for OCR AS Chemistry A, you will only need to know the content in the shaded box.

AS exam

Year 1 content
1. Development of practical skills in chemistry
2. Foundations in chemistry
3. Periodic table and energy
4. Core organic chemistry

Year 2 content
5. Physical chemistry and transition elements
6. Organic chemistry and analysis

A level exam

A Level exams will cover content from Year 1 and Year 2 and will be at a higher demand.

Contents

| How to use this book | v |

Module 2 Foundations in chemistry

Chapter 2 Atoms, ions, and compounds — 2
- 2.1 Atomic structure and isotopes — 2
- 2.2 Relative mass — 3
- 2.3 Compounds, formulae, and equations — 4
- Practice questions — 6

Chapter 3 Amount of substance — 7
- 3.1 Amount of substance and the mole — 7
- 3.2 Determination of formulae — 8
- 3.3 Moles and volumes — 10
- 3.4 Reacting quantities — 12
- Practice questions — 14

Chapter 4 Acids and redox — 15
- 4.1 Acids, bases, and neutralisation — 15
- 4.2 Acid–base titrations — 17
- 4.3 Redox — 19
- Practice questions — 21

Chapter 5 Electrons and bonding — 22
- 5.1 Electron structure — 22
- 5.2 Ionic bonding and structure — 24
- 5.3 Covalent bonding — 26
- Practice questions — 28

Chapter 6 Shapes of molecules and intermolecular forces — 29
- 6.1 Shapes of molecules and ions — 29
- 6.2 Electronegativity and polarity — 31
- 6.3 Intermolecular forces — 33
- 6.4 Hydrogen bonding — 34
- Practice questions — 35

Module 3 Periodic table and energy

Chapter 7 Periodicity — 36
- 7.1 The periodic table — 36
- 7.2 Ionisation energies — 37
- 7.3 Periodic trends in bonding and structure — 40
- Practice questions — 42

Chapter 8 Reactivity trends — 43
- 8.1 Group 2 — 43
- 8.2 The halogens — 45
- 8.3 Qualitative analysis — 47
- Practice questions — 48

Chapter 9 Enthalpy — 49
- 9.1 Enthalpy changes — 49
- 9.2 Measuring enthalpy changes — 51
- 9.3 Bond enthalpies — 53
- 9.4 Hess' law and enthalpy cycles — 55
- Practice questions — 57

Chapter 10 Reaction rates and equilibrium — 58
- 10.1 Reaction rates — 58
- 10.2 Catalysts — 60
- 10.3 The Boltzmann distribution — 61
- 10.4 Dynamic equilibrium and le Chatelier's principle — 63
- 10.5 The equilibrium constant, K_c — 65
- Practice questions — 66

Module 4 Core organic chemistry

Chapter 11 Basic concepts of organic chemistry — 67
- 11.1 Organic chemistry — 67
- 11.2 Nomenclature of organic compounds — 69
- 11.3 Representing the formulae of organic compounds — 71
- 11.4 Isomerism — 72
- 11.5 Introduction to reaction mechanisms — 73
- Practice questions — 74

Chapter 12 Alkanes — 75
- 12.1 Properties of the alkanes — 75
- 12.2 Chemical reactions of the alkanes — 76
- Practice questions — 78

Chapter 13 Alkenes — 79
- 13.1 The properties of alkenes — 79
- 13.2 Stereoisomerism — 80
- 13.3 Reactions of alkenes — 82
- 13.4 Electrophilic addition in alkenes — 83
- 13.5 Polymerisation in alkenes — 85
- Practice questions — 86

Chapter 14 Alcohols — 87
- 14.1 Properties of alcohols — 87
- 14.2 Reactions of alcohols — 88
- Practice questions — 90

Chapter 15 Haloalkanes — 91
- 15.1 The chemistry of the haloalkanes — 91
- 15.2 Organohalogen compounds in the environment — 93
- Practice questions — 94

Chapter 16 Organic synthesis — 95
- 16.1 Practical techniques in organic chemistry — 95
- 16.2 Synthetic routes — 96
- Practice questions — 98

Chapter 17 Spectroscopy — 99
- 17.1 Mass spectrometry — 99
- 17.2 Infrared spectroscopy — 101
- Practice questions — 104

Answers to summary questions	105
Answers to practice questions	114
Periodic table	120
Data sheets	121

How to use this book

This book contains many different features. Each feature is designed to support and develop the skills you will need for your examinations, as well as foster and stimulate your interest in chemistry.

 Worked example
Step-by-step worked solutions.

Summary Questions

1 These are short questions at the end of each topic.

2 They test your understanding of the topic and allow you to apply the knowledge and skills you have acquired.

3 The questions are ramped in order of difficulty. Lower-demand questions have a paler background, with the higher-demand questions having a darker background. Try to attempt every question you can, to help you achieve your best in the exams.

Chapter 2 Practice questions

Practice questions at the end of each chapter, including questions that cover practical and math skills.

1 Which row has the correct atomic structure for the $^{31}P^{3-}$ ion?

	protons	Neutrons	electrons
A	15	16	18
B	15	16	31
C	16	15	17
D	31	16	15

(1 mark)

2 What is the formula of sodium sulfate?
A NaS   B Na_2S
C $NaSO_4$ D Na_2SO_4 (1 mark)

3 What is the correct name for $Fe(NO_3)_3$?
A iron nitrate(III)   B iron(III) nitrite
C iron(III) nitrate   D iron nitride (1 mark)

4 What is the total number of electrons in a hydroxide ion?
A 9   B 10
C 11   D 12 (1 mark)

5 Lithium exists as a mixture of isotopes.
  a What is meant by the term *isotopes*? (1 mark)
  b The relative atomic mass of lithium can be calculated from the abundances of its isotopes.
    i Define the term *relative isotopic mass*. (2 marks)
    ii A sample of lithium consists of 7.6% ^6Li and 92.4% ^7Li by mass.
       Calculate the relative atomic mass of lithium in this sample. Give your answer to **two** decimal places. (2 marks)
  c In its reactions, different isotopes of lithium react in the same way.
    i Why do different isotopes of the same element react in the same way? (1 mark)
    ii Lithium reacts with water to form a solution of lithium hydroxide and hydrogen.
       Write the equation, including state symbols for this reaction. (2 marks)

6 A sample of gallium is analysed in a mass spectrometer to produce the mass spectrum.
  a Which isotope is used as the standard for measuring relative atomic masses? (1 mark)
  b i Estimate the percentage composition of each gallium isotope. (1 mark)
    ii Calculate the relative atomic mass of gallium in the sample. Give your answer to **one** decimal place. (2 marks)
  c Complete the table to show the number of sub-atomic particles in a ^{69}Ga atom and a $^{71}Ga^{3+}$ ion. (2 marks)

	protons	neutrons	electrons
^{69}Ga			
$^{71}Ga^{3+}$			

  d In its reactions, solid gallium forms compounds containing Ga^{3+} ions.
    i Gallium reacts with oxygen and with chlorine to form two solid compounds.
       Write equations, with state symbols, for the two reactions. (2 marks)
    ii Gallium reacts with dilute sulfuric acid to form a solution containing gallium sulfate and hydrogen.
       Write the equation, including state symbols for this reaction. (2 marks)

> **Specification references**
> → At the beginning of each topic, there are specification references to allow you to monitor your progress

> **Key term**
> Pulls out key terms for quick reference.

> **Revision tips**
> Revision tips contain prompts to help you with your understanding and revision.

> **Synoptic link**
> These highlight the key areas where topics relate to each other. As you go through your course, knowing how to link different areas of chemistry together becomes increasingly important. Many exam questions, particularly at A Level, will require you to bring together your knowledge from different areas.

2.1 Atomic structure and isotopes

Specification reference: 2.1.1

Atoms and sub-atomic particles

Atoms are made from sub-atomic particles called protons, neutrons, and electrons.

Protons and neutrons are located in the nucleus of the atom while electrons are found in a region outside of the nucleus in shells.

The mass of an electron is so small that it is often considered to be negligible.

▼ **Table 1** *The relative mass and relative charge of the sub-atomic particles*

Particle	Relative mass	Relative charge
Proton	1	+1
Neutron	1	0
Electron	1/1836	−1

Isotopes

Isotopes are atoms of the same element with different numbers of neutrons and different masses.

Representing isotopes

You need to use the terms: atomic number and mass number.

Atoms of an isotope of an element X can be represented: $^{A}_{Z}X$

- Number of protons = atomic number (Z)
- Number of electrons = atomic number (Z) (**only** for neutral atoms)
- Number of neutrons = mass number (A) − atomic number (Z) = $A - Z$

Different isotopes have the same atomic number but a different mass number.

There are two isotopes of chlorine:

$^{37}_{17}Cl$ protons = 17, electrons = 17, neutrons = (37 − 17) = 20

$^{35}_{17}Cl$ protons = 17, electrons = 17, neutrons = (35 − 17) = 18

For ions (particles which have lost or gained electrons), the ionic charge needs to be taken into account.

Beryllium atoms $^{9}_{4}Be$ protons = 4, electrons = 4, neutrons = 9 − 4 = 5

Fluoride ions $^{19}_{9}F^-$ protons = 9, electrons = 9 + 1 = 10, neutrons = 19 − 9 = 10

Calcium ions $^{42}_{20}Ca^{2+}$ protons = 20, electrons = 20 − 2 = 18, neutrons = 42 − 20 = 22

Properties of isotopes

The chemistry of an atom is determined by the behaviour of its electrons. All isotopes of an element have the same number of electrons and the same electron configuration. Therefore, different isotopes of an element have the same chemical reactions.

However, isotopes of an element have different numbers of neutrons and different masses. This may cause slight variations between isotopes in physical properties, such as boiling point and density.

> **Key term**
>
> **Isotopes:** Atoms of the same element with different numbers of neutrons and different masses. Isotopes have the same atomic number but a different mass number.

> **Key term**
>
> **Atomic number (Z):** The number of protons in the nucleus.

> **Key term**
>
> **Mass number (A):** The number of protons plus neutrons in the nucleus.

> **Summary questions**
>
> 1 State the relative charge and relative mass of the sub-atomic particles in the nucleus of an atom.
> *(2 marks)*
>
> 2 Explain why chlorine-35 and chlorine-37 are chemically identical.
> *(1 mark)*
>
> 3 Write down the numbers of protons, neutrons, and electrons in the following:
>   a $^{23}_{11}Na$ b $^{56}_{26}Fe^{3+}$
>   c $^{80}_{35}Br$ d $^{35}_{17}Cl^-$
> *(4 marks)*

2.2 Relative mass

Specification reference: 2.1.1

Relative isotopic mass

Relative masses are measured against the relative mass of carbon-12, which is set as exactly 12.

Relative isotopic mass is the mass of an atom of an isotope compared with one-twelfth of the mass of an atom of carbon-12. Sodium-23 has a relative isotopic mass of 23.0.

Relative atomic mass, A_r

Relative atomic mass takes into account the isotopes present in a sample of an element and the relative abundances of the isotopes.

Relative atomic mass, A_r, is the weighted mean mass of an atom of an element compared with 1/12 mass of carbon-12.

Magnesium has a relative atomic mass of 24.3.

Relative atomic mass can be calculated using a technique called mass spectrometry. A mass spectrum is obtained, showing the isotopes present in a sample of an element and the relative abundances of the isotopes.

- In a mass spectrum, the positive ions of isotopes are shown as a mass/charge ratio, m/z. Provided that the ionic charge is +1, m/z is the same as the mass of the ion.
- The relative abundances give the proportion of each isotope in the sample.

The mass spectrum of a sample of lead is shown in Figure 1.

The A_r of lead can be calculated as a weighted mean mass:

$$A_r = \frac{(204 \times 1.4) + (206 \times 24.1) + (207 \times 22.1) + (208 \times 52.4)}{100} = 207.24$$

Relative molecular mass and relative formula mass

Molecular formulae are used for simple molecular compounds such as ethene, C_2H_4. The relative molecular mass of ethene is found by adding the relative atomic masses of all the atoms in the molecular formula of ethene.

$$M_r \text{ of } C_2H_4 = (12.0 \times 2) + (1.0 \times 4) = 28.0$$

Relative formula mass is used for giant ionic compounds such as potassium chloride KCl, and giant covalent compounds such as silicon dioxide SiO_2.

relative formula mass of $MgCl_2 = 24.3 + (35.5 \times 2) = 95.3$

relative formula mass of $Fe(NO_3)_2 = 55.8 + (14.0 + 16.0 \times 3) \times 2 = 179.8$

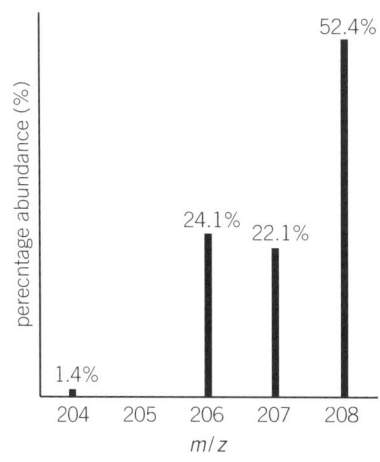

▲ Figure 1 *The mass spectrum of lead*

Revision tip
The relative isotopic mass of the carbon-12 isotope is exactly 12. One-twelfth of exactly 12 is exactly 1.

Carbon-12 is the international standard for measuring all relative masses.

Synoptic link
You will find out more about mass spectrometry in Topic 17.1, Mass spectrometry.

Synoptic link
For more details of molecular formulae, see Topic 3.2, Determination of formulae.

Synoptic link
For more details of giant structures, see Topic 5.2, Ionic bonding and structure, and Topic 7.3, Periodic trends in bonding and structure.

Key terms
Relative isotopic mass: The mass of an isotope compared with 1/12 mass of carbon-12.

Relative atomic mass: The weighted mean mass of an atom compared with 1/12 mass of carbon-12.

Summary questions

1. Calculate the relative molecular mass of the following:
   a C_5H_{12} b H_3PO_4 c C_2H_5OH *(3 marks)*

2. Calculate the relative formula mass of the following:
   a K_2O b $NiCO_3$ c $(NH_4)_2SO_4$ *(3 marks)*

3. Use the mass spectrum data of neon below to determine the relative atomic mass of neon to **two** decimal places. *(2 marks)*

m/z	20.0	21.0	22.0
relative abundance (%)	90.90	0.300	8.80

4. A sample of lithium was analysed and found to contain two isotopes. The sample has a relative atomic mass of 6.922 and contains 7.80% lithium-6. What is the other isotope present in the sample? *(2 marks)*

2.3 Compounds, formulae, and equations

Specification reference: 2.1.2

> **Synoptic link**
>
> For details of empirical formulae, see Topic 3.2, Determination of formulae.
>
> For details of ionic bonding and an ionic lattice, see Topic 5.2, Ionic bonding and structure.

Ionic formulae

You cannot write a molecular formula for an ionic compound. An empirical formula is written instead, showing the ratio of cations (positive ions) to anions (negative ions) present in the ionic lattice. The formula for an ionic compound is worked out from the ionic charges of the cations and anions present.

Ionic charges

Ions from the periodic table

You can often work out an ionic charge from the position of the element in the periodic table.

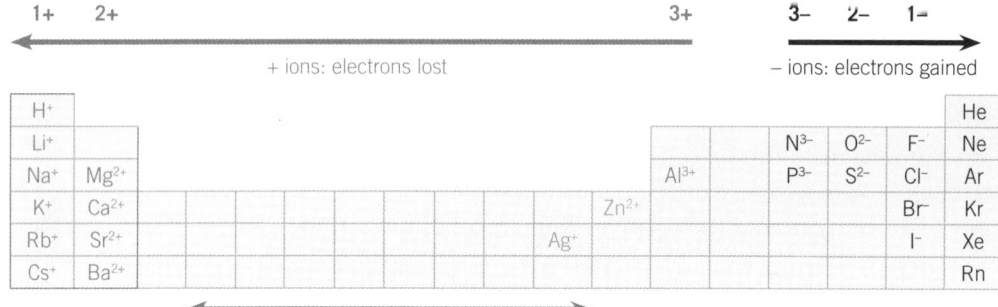

▲ **Figure 1** *Ionic charges and the periodic table*

Ions of transition metals

Transition metals can form ions with different charges. You can work out the charge from the name of the transition metal compound.

- Iron(II) oxide contains Fe^{2+} ions.
- Iron(III) oxide contains Fe^{3+} ions.

Writing an ionic formula

Ionic compounds have a neutral charge overall. This means that:

- total positive charge from cations = total negative charge from anions.

For example:

- Barium chloride contains Ba^{2+} ions and Cl^- ions and has the formula $BaCl_2$. Overall charge: $(2+) + (1- \times 2) = 0$

Some ionic compounds contain polyatomic ions.

For example:

- Aluminium sulfate contains Al^{3+} ions and SO_4^{2-} ions and has the formula $Al_2(SO_4)_3$. Overall charge $= (3+ \times 2) + (2- \times 3) = 0$

Ions to learn

You will need to recall the names and ionic charges in Table 1. These cannot be worked out directly from the periodic table.

> **Key terms**
>
> **Cation:** A positively charged ion, e.g. Na^+.
>
> **Anion:** A negatively charged ion, e.g. O^{2-}.

> **Synoptic link**
>
> You will find out more about using Roman numerals within names in Topic 4.3, Redox.

▼ **Table 1**

Negative ions	Positive ions
Nitrate, NO_3^-	Ammonium, NH_4^+
Carbonate, CO_3^{2-}	Zinc, Zn^{2+}
Sulfate, SO_4^{2-}	Silver, Ag^+
Hydroxide, OH^-	

Balancing chemical equations

To write and balance an equation:

- Write the formulae of the reactants and products in an unbalanced equation.
- Multiply each formula by a 'balancing number', in front of the formula, until the number of atoms of each element is the same on each side of the equation.
- The equation is balanced when all elements have the same number of atoms on each side of the equation.

Atoms, ions, and compounds

 Worked example: Reaction of sodium and chlorine

Sodium reacts with chlorine to form sodium chloride.
- Write the formulae of the reactants and products in an unbalanced equation.

$$Na + Cl_2 \rightarrow NaCl$$

- Balance the equation by placing numbers in front of the formulae.

$$2Na + Cl_2 \rightarrow 2NaCl$$

Left-hand side has **2** Na and **2** Cl

Right-hand side has **2** × NaCl = **2** Na and **2** Cl

The equation is balanced.

- Finally, add state symbols to complete the equation.

$$2Na(s) + Cl_2(g) \rightarrow 2NaCl(s)$$

Ionic equations

Ionic equations can be written for many reactions involving ions.

 Worked example: Writing an ionic equation for precipitation

A precipitation reaction occurs when two solutions containing ions are mixed to form a precipitate.

Precipitation reactions are often used to test for the presence of ions. For example, when silver nitrate solution is mixed with potassium iodide solution, a yellow precipitate of silver iodide is formed.

The full equation is: $AgNO_3(aq) + KI(aq) \rightarrow KNO_3(aq) + AgI(s)$

All the ions are written out and common ions on both sides of the equation are cancelled:

$Ag^+(aq) + \cancel{NO_3^-(aq)} + \cancel{K^+(aq)} + I^-(aq) \rightarrow \cancel{K^+(aq)} + \cancel{NO_3^-(aq)} + AgI(s)$

The ionic equation is: $Ag^+(aq) + I^-(aq) \rightarrow AgI(s)$

Revision tip

When balancing an equation, you must **not** change any chemical formula.

It may be tempting to balance by changing the formula:

$Na + Cl_2 \rightarrow NaCl_2$ ✗

This is a serious error.

If you change the formula to $NaCl_2$, this is not sodium chloride anymore!

▼ **Table 2** *State symbols*

State symbol	Meaning
(s)	Solid
(l)	Liquid
(g)	Gas
(aq)	Aqueous/ dissolved in water

Synoptic link

For details of tests for ions, see Topic 8.3, Qualitative analysis.

Revision tip

Ions that play no part in the reaction are called spectator ions. Spectator ions do not change and will be on both sides of the equation. These are cancelled when writing the ionic equation.

Summary questions

1. Write the formula for the following ionic compounds:
    a. sodium oxide
    b. silver carbonate
    c. calcium hydroxide
    d. ammonium carbonate
    e. chromium(III) sulfate. *(5 marks)*

2. Balance the following equations:
    a. $KClO_3(s) \rightarrow KCl(s) + KClO_4(s)$ *(1 mark)*
    b. $Pb(NO_3)_2(s) \rightarrow PbO(s) + NO_2(g) + O_2(g)$ *(1 mark)*
    c. $NH_3(g) + O_2(g) \rightarrow NO(g) + H_2O(l)$ *(1 mark)*

3. Write balanced equations, with state symbols, for the following reactions:
    a. Calcium carbonate solid reacts with dilute hydrochloric acid to form calcium chloride solution, carbon dioxide gas, and water. *(1 mark)*
    b. Aluminium reacts with copper(II) nitrate solution to form copper and aluminium nitrate solution. *(1 mark)*
    c. Concentrated hydrochloric acid reacts with manganese(IV) oxide solid to form a solution of manganese(II) chloride, chlorine gas, and water. *(1 mark)*

5

Chapter 2 Practice questions

1. Which row has the correct atomic structure for the $^{31}P^{3-}$ ion?

	protons	Neutrons	electrons
A	15	16	18
B	15	16	31
C	16	15	17
D	31	16	15

   (1 mark)

2. What is the formula of sodium sulfate?

   A NaS B Na_2S
   C $NaSO_4$ D Na_2SO_4 *(1 mark)*

3. What is the correct name for $Fe(NO_3)_3$?

   A iron nitrate(III) B iron(III) nitrite
   C iron(III) nitrate D iron nitride *(1 mark)*

4. What is the total number of electrons in a hydroxide ion?

   A 9 B 10
   C 11 D 12 *(1 mark)*

5. Lithium exists as a mixture of isotopes.

   a What is meant by the term *isotopes*? *(1 mark)*

   b The relative atomic mass of lithium can be calculated from the abundances of its isotopes.

   i Define the term *relative isotopic mass*. *(2 marks)*

   ii A sample of lithium consists of 7.6% ^{6}Li and 92.4% ^{7}Li by mass.

   Calculate the relative atomic mass of lithium in this sample. Give your answer to **two** decimal places. *(2 marks)*

   c In its reactions, different isotopes of lithium react in the same way.

   i Why do different isotopes of the same element react in the same way? *(1 mark)*

   ii Lithium reacts with water to form a solution of lithium hydroxide and hydrogen.

   Write the equation, including state symbols for this reaction. *(2 marks)*

6. A sample of gallium is analysed in a mass spectrometer to produce the mass spectrum.

   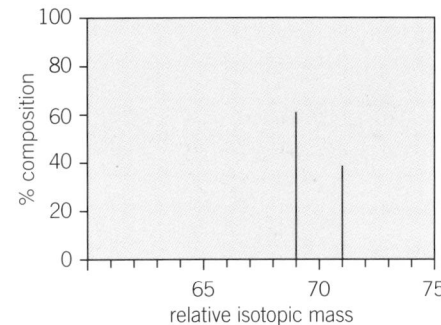

   a Which isotope is used as the standard for measuring relative atomic masses? *(1 mark)*

   b i Estimate the percentage composition of each gallium isotope. *(1 mark)*

   ii Calculate the relative atomic mass of gallium in the sample. Give your answer to **one** decimal place. *(2 marks)*

   c Complete the table to show the number of sub-atomic particles in a ^{69}Ga atom and a $^{71}Ga^{3+}$ ion. *(2 marks)*

	protons	neutrons	electrons
^{69}Ga			
$^{71}Ga^{3+}$			

   d In its reactions, solid gallium forms compounds containing Ga^{3+} ions.

   i Gallium reacts with oxygen and with chlorine to form two solid compounds.

   Write equations, with state symbols, for the two reactions. *(2 marks)*

   ii Gallium reacts with dilute sulfuric acid to form a solution containing gallium sulfate and hydrogen.

   Write the equation, including state symbols for this reaction. *(2 marks)*

3.1 Amount of substance and the mole
Specification reference: 2.1.3

The mole and the Avogadro constant, N_A
- The mole (symbol: mol) is the unit for amount of substance.
- One mole of a substance contains the same number of particles as there are atoms in exactly 12 g of ^{12}C.
- The Avogadro constant N_A is 6.02×10^{23} mol^{-1}, the number of particles per mole of particles.

Molar mass, M
Molar mass (in g mol^{-1}) is the mass in grams of one mole of substance.

Ethene, C_2H_4, has a molar mass of $(12.0 \times 2 + 1.0 \times 4) = 28.0$ g mol^{-1}.

Amount of substance
Amount of substance n (in mol) is calculated from the mass m (in g) and molar mass M (in g mol^{-1}) using the equation:

$$\text{Amount, } n = \frac{\text{mass, } m}{\text{molar mass, } M} \text{ or } n = \frac{m}{M}$$

You must be able to rearrange this equation so that you can calculate any one of these three quantities from the other two.

> **Worked example: Mole calculations using mass and molar mass**
>
> **Q1** Calculate the amount, in mol, of Mg in 48.6 g Mg.
> $$n = \frac{m}{M} = \frac{48.6}{24.3} = 2.00 \text{ mol}$$
>
> **Q2** Calculate the mass, in g, of 0.740 mol of $Ca(OH)_2$.
> $$n = \frac{m}{M} \quad \therefore \text{mass } m = n \times M = 0.740 \times 74.1 = 54.8 \text{ g}$$
>
> **Q3** 0.500 mol of a compound has a mass of 22.0 g. Calculate the molar mass.
> $$n = \frac{m}{M} \quad \therefore \text{molar mass } M = \frac{m}{n} = \frac{22.0}{0.500} = 44.0 \text{ g mol}^{-1}$$

Number of particles
The number of particles can be calculated by multiplying the amount in moles by the Avogadro constant: i.e. number of particles = $n \times 6.02 \times 10^{23}$.

Summary questions

1. How many atoms are in the following.
   a 0.300 mol of Na *(1 mark)*
   b 0.750 mol of N_2 *(1 mark)*

2. Calculate the following, to **three** significant figures:
   a the amount, in mol, of CaO in 5.61 g of CaO *(1 mark)*
   b the mass of 0.650 mol of P_4 *(1 mark)*
   c the mass of 0.350 mol of $Al(OH)_3$. *(1 mark)*

3. Calculate the following, to **three** significant figures:
   a the amount, in mol, of H_3BO_3 in 12.36 g of H_3BO_3 *(1 mark)*
   b the mass of 0.150 mol of $(NH_4)_2CO_3$ *(1 mark)*
   c the mass of 4.515×10^{24} molecules of PCl_3. *(2 marks)*

> **Synoptic link**
> You learnt about carbon-12 in Topic 2.2, Relative mass.

> **Revision tip**
> The value of the Avogadro constant is given on the data sheet and does not have to be memorised.

> **Revision tip**
> The numerical value of molar mass is the same as relative atomic mass, relative molecular mass, or relative formula mass.

> **Revision tip**
> Remember, remember, remember:
> $$\text{Amount, } n = \frac{\text{mass, } m}{\text{molar mass, } M}$$
> You will use this equation many times to work out any of the three variables, n, m or M.
>
> Make sure you can rearrange this important equation to give:
>
> Amount $\quad n = \frac{m}{M}$
>
> Mass $\quad m = n \times M$
>
> Molar mass $\quad M = \frac{m}{n}$

> **Worked example: Calculating number of particles**
>
> **Q1** Calculate the number of Al atoms in 0.500 mol of Al.
>
> Number of Al atoms = $0.500 \times 6.02 \times 10^{23} = 3.01 \times 10^{23}$
>
> **Q2** Calculate the number of O_2 molecules and O atoms in 2.00 mol of O_2.
>
> Number of O_2 molecules = $2.00 \times 6.02 \times 10^{23} = 12.04 \times 10^{23}$
>
> Each O_2 molecule contains 2 O atoms.
>
> $\therefore$ number of O atoms = $2 \times 12.04 \times 10^{23} = 24.08 \times 10^{23}$

3.2 Determination of formulae
Specification reference: 2.1.1, 2.1.3

Synoptic link
Empirical formula and molecular formula are very important in organic chemistry. See Topic 11.3, Representing the formulae of organic compounds.

Revision tip
In empirical formula calculations, you may be provided with percentage compositions, by mass, or the masses of the elements. Both calculations are answered in the same way. See the worked examples.

Revision tip
It is essential to show all your working in these calculations; $\div A_r$ gives the ratio of the moles of atoms.

Revision tip
Sometimes, you may not be provided with the mass of one of the elements.

You can easily find the mass if you know the total mass and the masses of the other elements present. See the worked example.

Key terms
Empirical formula: The simplest whole number ratio of atoms of each element present in a compound.

Molecular formula: The number and type of atoms of each element in a molecule.

Empirical and molecular formulae

Empirical formula is the simplest whole number ratio of atoms of each element present in a compound.

Molecular formula is the number and type of atoms of each element in a molecule.

For example: The empirical formula of butane is C_2H_5.

The molecular formula of butane is C_4H_{10}.

Calculating an empirical formula

The empirical formula of a compound can be calculated from the percentage composition, by mass, or mass of the elements in the compound.

Worked example: Calculating an empirical formula

A compound contains 40.06% calcium, 11.99% carbon, and 47.95% oxygen by mass.

Determine the empirical formula of the compound.

Element	Ca	C	O
% by mass	40.06	11.99	47.95
$\div A_r$	0.999	0.999	2.997
$\div$ smallest (0.999)	1	1	3

Empirical formula = $CaCO_3$

Calculating a molecular formula

The molecular formula of a compound can be found if the empirical formula and the molecular mass are known.

Worked example: Calculating empirical and molecular formulae

A hydrocarbon has a relative molecular mass of 84.0. A 4.20 g sample of the hydrocarbon was analysed and found to contain 3.60 g of carbon. Calculate the molecular formula.

Step 1: Calculate the empirical formula.

The compound is a hydrocarbon and contains carbon and hydrogen only:

Mass of C = 3.60 g   Mass of H = 4.20 − 3.60 = 0.60 g

Element	C	H
mass/g	3.60	0.60
$\div A_r$	0.30	0.60
$\div$ smallest (0.30)	1	2

Empirical formula = CH_2

Step 2: Calculate the molecular formula.

The relative molecular mass of the compound is 84.0

The empirical formula is CH_2 which has a mass of (12.0 + 2.0) = 14.0

Empirical formula units = $\dfrac{84.0}{14.0} = 6$

The molecular formula is C_6H_{12}

Amount of substance

Hydrated salts

- A hydrated compound is crystalline and contains water molecules.
- An anhydrous compound contains no water molecules.
- Water of crystallisation refers to water molecules that are bonded into the crystalline structure of a hydrated compound. In the formula, the amount of water is shown after a dot (•).

For example:

$CuSO_4(s)$ anhydrous copper(II) sulfate

$CuSO_4 \cdot 5H_2O(s)$ hydrated copper(II) sulfate

Hydrated copper(II) sulfate contains 5 moles of water of crystallisation per mole of the hydrated salt.

Calculating the formula of a hydrated salt

The formula of a hydrated salt can be calculated from the results of an experiment, as shown below.

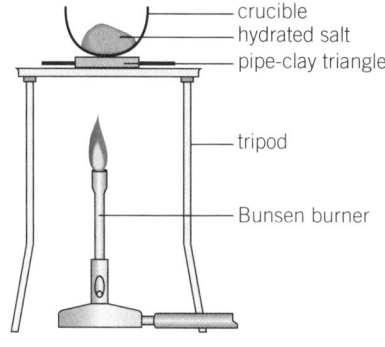

▲ Figure 1 *Apparatus for removing water from a hydrated salt*

Worked example: Calculating the formula of a hydrated salt

Hydrated sodium sulfate, $Na_2SO_4 \cdot xH_2O$ is heated to remove the water of crystallisation.

Mass of hydrated sodium sulfate before heating = 3.221 g

Mass of anhydrous sodium sulfate after heating = 1.421 g

Determine the formula of hydrated sodium sulfate.

Step 1: Calculate the amount, in mol, of anhydrous Na_2SO_4 and H_2O.

Mass of H_2O removed = 3.221 − 1.421 = 1.800 g

	Na_2SO_4		H_2O
mass in g	1.421		1.800
Amount in mol	$\frac{1.421}{142.1}$	:	$\frac{1.800}{18.0}$
	= 0.0100	:	0.100

Step 2: Calculate the ratio of the amounts and the formula.

	Na_2SO_4		H_2O
	0.0100	:	0.100
÷ smallest (0.01)	$\frac{0.0100}{0.0100}$	:	$\frac{0.100}{0.0100}$
	= 1	:	10

Formula = $Na_2SO_4 \cdot 10H_2O$

Revision tip

You can check that the water of crystallisation has been removed by **heating to constant mass**:

- The hydrated salt is heated and weighed.
- The salt is heated again and reweighed.
- Continue until the mass no longer changes.

You can then assume that all the water of crystallisation has been removed.

Summary questions

1. The formula of hydrated cobalt(II) chloride is $CoCl_2 \cdot 6H_2O$.
   a. What does the term 'hydrated' mean? *(1 mark)*
   b. What does the '$6H_2O$' part of the formula show? *(1 mark)*

2. A 0.399 g sample of an iron oxide was analysed and found to contain 0.279 g of iron.
   Calculate the empirical formula of this compound. *(2 marks)*

3. 6.45 g of hydrated magnesium sulfate, $MgSO_4 \cdot xH_2O$, were heated until all the water of crystallisation had been removed. 3.16 g of anhydrous magnesium sulfate remained.
   Determine the formula of the hydrated magnesium sulfate. *(3 marks)*

3.3 Moles and volumes
Specification reference: 2.1.3

Concentrations

Many chemical experiments are carried out in aqueous solution.

- Concentration is the amount (in mol) of a dissolved substance in $1.00\,dm^3$ of solution ($1\,dm^3 = 1000\,cm^3$). Concentration has units of $mol\,dm^{-3}$.
- For example, a $2.00\,mol\,dm^{-3}$ solution of sulfuric acid contains $2.00\,mol$ H_2SO_4 dissolved in $1.00\,dm^3$ of solution.

The relationship between amount n (in mol), concentration c (in $mol\,dm^{-3}$) and volume V (in dm^3) is:

$$\text{Amount, } n = \text{concentration, } c \times \text{volume, } V \quad \text{or} \quad n = c \times V$$

Other ways of measuring concentration

Concentration can also be expressed as the mass of a dissolved substance in grams in $1\,dm^3$ of solution. This 'mass concentration' has units of $g\,dm^{-3}$.

$2.00\,mol\,dm^{-3}$ H_2SO_4 has a mass concentration of H_2SO_4 of $2.00 \times 98.1 = 196.2\,g\,mol^{-1}$.

> **Revision tip**
> **Moles and concentrations**
>
> If V is in dm^3: $n = c \times V$
>
> You will use this equation many times to work out any of the three variables, n, c or V.
>
> Make sure you can rearrange and use this important equation:
>
> Amount: $n = c \times V$
>
> Concentration: $c = \dfrac{n}{V}$
>
> Volume: $V = \dfrac{n}{c}$
>
> If V is in cm^3, you will need to divide V by 1000 to convert to dm^3.
>
> Amount: $n = c \times \dfrac{V}{1000}$
>
> Rearranging gives:
>
> Concentration: $c = n \times \dfrac{1000}{V}$
>
> Volume: $V = \dfrac{n \times 1000}{c}$

Worked example: Concentration calculations

Q1 Calculate the amount, in mol, of HCl in $50.0\,cm^3$ of $0.100\,mol\,dm^{-3}$ solution.

$$\text{Amount } n = c \times \dfrac{V}{1000} = 0.100 \times \dfrac{50.0}{1000} = 0.005\,00\,mol$$

Q2 Calculate the volume, in cm^3, of $0.0500\,mol\,dm^{-3}$ NaOH that contains $1.15 \times 10^{-3}\,mol$ of NaOH.

$$\text{Volume } V = \dfrac{n \times 1000}{c} = \dfrac{1.15 \times 10^{-3} \times 1000}{0.0500} = 23.0\,cm^3$$

Q3 $50.0\,cm^3$ of HNO_3 contains $1.25 \times 10^{-3}\,mol$. Calculate the concentration of the HNO_3.

$$\text{Concentration } c = n \times \dfrac{1000}{V} = 1.25 \times 10^{-3} \times \dfrac{1000}{50.0} = 0.0250\,mol\,dm^{-3}$$

Molar gas volume

Molar gas volume is the volume per mole of a gas at a stated temperature and pressure.

- Molar gas volume increases as temperature increases.
- Molar gas volume decreases as pressure increases.

At room temperature and pressure (RTP), $1.00\,mol$ of any gas occupies a volume of $24.0\,dm^3$ and has a molar gas volume of $24.0\,dm^3\,mol^{-1}$.

> **Revision tip**
> The value for the molar gas volume at RTP is given on the data sheet and does not have to be memorised.

Worked example: Calculating moles from gas volume at RTP

At RTP, a sample of gas molecules occupies $18.0\,dm^3$.

Calculate the amount in mol of gas molecules present in this sample.

$$\text{Amount} = \dfrac{18.0}{24.0} = 0.750\,mol$$

> **Revision tip**
> The gas does not matter. One mole of any gas occupies $24.0\,dm^3$ at RTP. This applies to $2\,g\,H_2$, $28\,g\,N_2$, $32\,g\,O_2$, and $44\,g\,CO_2$.
>
> Although the number of moles and volume are the same, the masses are different and the gases have different densities.

Worked example: Calculating gas volume at RTP from moles

What is the volume, in cm^3, of $0.004\,25\,mol$ of O_2 gas at room temperature and pressure?

$$1.00\,mol \text{ of } O_2 \text{ gas occupies } 24.0\,dm^3 = 24\,000\,cm^3$$

$$0.004\,25\,mol \text{ of } O_2 \text{ gas occupies } 0.004\,25 \times 24\,000 = 102\,cm^3$$

The ideal gas equation

The ideal gas equation links the volume of an amount of a gas with pressure and temperature.

Ideal gas equation: $pV = nRT$

p: pressure (in Pa) V: volume (in m³) T: temperature (in K)

n: amount (in mol) R: gas constant = $8.314\,\mathrm{J\,mol^{-1}\,K^{-1}}$

 Worked example

Finding the pressure of a gas
What is the pressure of 0.200 mol of chlorine gas in a container of volume 6.00 m³ at a temperature of 298 K?

$$pV = nRT \qquad \therefore p = \frac{nRT}{V} = \frac{0.200 \times 8.314 \times 298}{6.00} = 82.6\,\mathrm{Pa}$$

Finding the volume of a gas
What is the volume, in cm³, of 6.40×10^{-4} mol of oxygen gas at a pressure of 100 kPa at 0 °C?

$$pV = nRT \qquad \therefore V = \frac{nRT}{p} = \frac{6.40 \times 10^{-4} \times 8.314 \times 273}{100 \times 10^{3}} = 1.45 \times 10^{-5}\,\mathrm{m^{3}} = 14.5\,\mathrm{cm^{3}}$$

Finding the temperature of a gas
At what temperature would 10.0 mol of hydrogen gas occupy a volume of 24.4 dm³ at a pressure of 200 kPa?

$$pV = nRT \qquad \therefore T = \frac{pV}{nR} = \frac{200 \times 10^{3} \times 24.4 \times 10^{-3}}{10.0 \times 8.314} = 58.7\,\mathrm{K}$$

Determination of relative molecular mass, M_r

The ideal gas equation can be used in determine the relative molecular gas of a volatile liquid.

- The liquid is weighed and then heated to a known temperature above its boiling point.
- The volume of the gas is measured and the pressure recorded.

 Worked example: Calculating relative molecular mas from the ideal gas equation

At 100 °C and 100 kPa pressure, 0.2025 g of a gas occupies a volume of 87.2 cm³.

What is the relative molecular mass of the gas?

Step 1: Ensure all the quantities are in the correct units: p in Pa, V in m³ and T in K.

Pressure, p 100 kPa = 100×10^{3} Pa

Volume, V 87.2 cm³ = 87.2×10^{-6} m³

Temperature, T: 100 °C = 100 + 273 = 373 K

Step 2: Rearrange the ideal gas equation and calculate the amount of gas, n, in mol.

$$pV = nRT \qquad \therefore n = \frac{pV}{RT} = \frac{100 \times 10^{3} \times 87.2 \times 10^{-6}}{8.314 \times 373} = 2.812 \times 10^{-3}\,\mathrm{mol}$$

Step 3: Calculate the relative molecular mass of the gas.

$$n = \frac{m}{M} \qquad \therefore \text{molar mass } M = \frac{m}{n} = \frac{0.2025}{2.812 \times 10^{-3}} = 72.0\,\mathrm{g\,mol^{-1}}$$

Relative molecular mass = 72.0

Amount of substance

Revision tip
The value for the gas constant, R, is given on the data sheet and does not need to be memorised.

Revision tip
When using the ideal gas equation, it is essential that you convert pressure, volume and temperature into the correct units:

p in Pa; V in m³, T in K.

Pressure conversions

1 kPa = 10^{3} Pa

kPa → Pa: multiply by 10^{3}

Pa → kPa: multiply by 10^{-3}

Volume conversions

1 cm³ = 10^{-6} m³

cm³ → m³ multiply by 10^{-6}

m³ → cm³ multiply by 10^{6}

Also: 1 cm³ = 10^{-3} dm³

         1 dm³ = 10^{-3} m³

Temperature conversions

0 °C = 273 K

°C → K: add 273

K → °C: subtract 273

Summary questions

1. Calculate the amount, in mol, of HNO_3 in 75.0 cm³ of 0.0500 mol dm⁻³ HNO_3. *(1 mark)*

2. A sample of gas at room temperature and pressure occupies 22.0 dm³. Calculate the amount, in mol, of gas present in this sample. *(1 mark)*

3. Calculate the volume occupied by 2.31 mol of hydrogen gas at a pressure of 200 kPa and a temperature of 32 °C. *(3 marks)*

3.4 Reacting quantities

Specification reference: 2.1.3

Stoichiometry

A balanced equation is used to show what happens in a chemical reaction. The balancing numbers show the ratio of the moles involved in the chemical reaction. This ratio is called the stoichiometry.

Example: $H_2 + Cl_2 \rightarrow 2HCl$

1 mol H_2 reacts with 1 mol Cl_2 to produce 2 mol HCl.

Quantities from equations

The stoichiometry of a balanced equation can be used to calculate the mass of products made in a reaction or the mass of reactants used in a reaction.

> ### Worked example: Using an equation to calculate reacting quantities
>
> 4.812 g of calcium is completely reacted with nitrogen to form calcium nitride, Ca_3N_2:
>
> $$3Ca(s) + N_2(g) \rightarrow Ca_3N_2(s)$$
>
> Calculate the mass of Ca_3N_2 formed and the volume of N_2, in cm^3, at RTP that reacts.
>
> **Step 1:** Calculate the amount, in moles, of Ca that reacts.
>
> $$n(Ca) = \frac{m}{M} = \frac{4.812}{40.1} = 0.120 \text{ mol}$$
>
> **Step 2:** Use the equation to find the reacting quantities in moles.
>
> From the balancing numbers in the equation: $3Ca(s) + N_2(g) \rightarrow Ca_3N_2(s)$
>
> 3 mol Ca reacts with   1 mol N_2 to form   1 mol Ca_3N_2
>
> ∴ 0.120 mol Ca reacts with 0.0400 mol N_2 to form 0.0400 mol Ca_3N_2
>
> **Step 3:** Calculate the mass of Ca_3N_2 formed and the volume of N_2 that reacts.
>
> $n(Ca_3N_2) = \frac{m}{M}$ ∴ $m = n \times M = 0.0400 \times (40.1 \times 3 + 14.0 \times 2)$
>
> mass m of $Ca_3N_2 = 0.0400 \times 148.3 = 5.932$ g
>
> 1 mol N_2 has a volume of 24.0 dm^3 at RTP
>
> ∴ 0.0400 mol N_2 has a volume of $0.0400 \times 24.0 = 0.96$ $dm^3 = 960$ cm^3

Synoptic link

See Topic 3.1, Amount of substance and the mole, for the relationship between amount in mol, mass, and molar mass:

$n = m/M$

Synoptic link

See Topic 3.3, Moles and volumes, for the relationship between amount in moles and gas volumes.

Atom economy

The atom economy of a chemical reaction is the percentage proportion of reactants that are converted into useful products:

$$\text{Atom economy} = \frac{\text{sum of molar masses of desired products}}{\text{sum of molar masses of all products}} \times 100\%$$

- A high atom economy is efficient and produce little waste.
- Sustainability favours reactions with a high atom economy: fewer raw materials are used and less waste is produced.

Amount of substance

 Worked example: Calculating atom economy

The reaction of methane with steam is used to produce hydrogen gas. Carbon monoxide is also formed as a waste product.

Calculate the atom economy of the reaction for producing hydrogen.

$$CH_4(g) + H_2O(g) \rightarrow 3H_2(g) + CO(g)$$

$$\text{Atom economy} = \frac{\text{sum of molar masses of desired products}}{\text{sum of molar masses of all products}} \times 100$$

$$= \frac{3 \times 2}{(3 \times 2) + 28} \times 100 = \frac{6}{34} \times 100 = 17.6\%$$

Revision tip
Notice that the balancing numbers have been taken into account as 3 mol H_2 is produced.

Percentage yield

Percentage yield is the actual yield of a product shown as a percentage of the theoretical yield:

$$\text{Percentage yield} = \frac{\text{actual yield}}{\text{theoretical yield}} \times 100\%$$

 Worked example: Calculating percentage yield

2.00 g of calcium carbonate decomposes on heating, to form 1.00 g of calcium oxide.

The equation is: $CaCO_3(s) \rightarrow CaO(s) + CO_2(g)$

What is the percentage yield of calcium oxide to **two** significant figures?

Step 1: Calculate the amount, in moles, of $CaCO_3$ that reacts:

$$n(CaCO_3) = \frac{m}{M} = \frac{2.00}{100.1} = 0.0200 \text{ mol}$$

Step 2: Use the equation to find the theoretical yield of CaO, in moles.

From the balancing numbers in the equation,

1 mol $CaCO_3$ forms 1 mol CaO

∴ 0.0200 mol $CaCO_3$ forms 0.0200 mol CaO

Step 3: Calculate the actual yield of CaO, in moles.

$$n(CaO) = \frac{m}{M} = \frac{1.00}{56.1} = 0.0178 \text{ mol}$$

Step 4: Calculate the percentage yield of CaO.

$$\% \text{ yield} = \frac{\text{actual yield}}{\text{theoretical yield}} \times 100\% = \frac{0.0178}{0.0200} \times 100 = 89\%$$

Revision tip
Reactions can have a high atom economy and a low percentage yield or a low atom economy and a high percentage yield.

Summary questions

1. Chlorine gas can be obtained from the electrolysis of brine.
   $2NaCl(aq) + 2H_2O(l) \rightarrow 2NaOH(aq) + Cl_2(g) + H_2(g)$
   Calculate the atom economy for producing chlorine. *(2 marks)*

2. Sodium reacts with oxygen to make sodium oxide, Na_2O.
   $4Na(s) + O_2(g) \rightarrow 2Na_2O(s)$
   Calculate the mass of Na required to produce 1.24 g of Na_2O. *(2 mark)*

3. 2.00 g of ethene, C_2H_4, reacts with hydrogen bromide, HBr, to form 5.80 g C_2H_5Br. Calculate the percentage yield for this reaction. *(3 marks)*

4. Magnesium metal reacts with hydrochloric acid to form hydrogen gas.
   $Mg(s) + 2HCl(aq) \rightarrow MgCl_2(aq) + H_2(g)$
   Calculate the volume, in cm^3, of gas released when 0.200 g of magnesium reacts with an excess of acid at 25 °C and 100 kPa. *(4 marks)*

Chapter 3 Practice questions

1. An oxide of vanadium and oxygen contains 56.04% by mass of vanadium. What is the empirical formula of the compound?

   A V_2O_3 **B** V_2O_5

   C V_3O_2 **D** V_5O_2 *(1 mark)*

2. A solution is prepared by dissolving 12.5 g of NaCl in enough water to make 350 cm³ of solution.

   What is the correct concentration, in mol dm⁻³, of the solution?

   A 6.11×10^{-4} **B** 3.57×10^{-2}

   C 6.11×10^{-1} **D** 35.7 *(1 mark)*

3. 1.155 g of a gas occupies 630 cm³ at room temperature and pressure.

   What is the relative molecular mass of the gas?

   A 27.7   **B** 30.3

   C 44.0   **D** 545.5 *(1 mark)*

4. A student reacts an excess of strontium carbonate with 64.0 cm³ of 0.800 mol dm⁻³ hydrochloric acid. The equation is shown below.

   $$SrCO_3(s) + 2HCl(aq) \rightarrow SrCl_2(aq) + H_2O(l) + CO_2(g)$$

   a. Calculate the minimum mass of $SrCO_3$ needed to completely react with the HCl.

   Give your answer to **three** significant figures. *(3 marks)*

   b. The student removed the unreacted $SrCO_3$ and made up the resulting solution of $SrCl_2$ to 500 cm³ in a volumetric flask.

   i. How could the student have removed the unreacted $SrCO_3$? *(1 mark)*

   ii. Calculate the concentration of $SrCl_2$ obtained in the volumetric flask. *(1 mark)*

   c. A compound of strontium has the percentage composition by mass: Sr, 48.78%; N, 15.59%; O, 35.63%.

   Calculate the empirical formula of the compound. *(2 marks)*

5. This question looks at different gases.

   a. A 0.222 g sample of a gas occupies 198 cm³ at a pressure of 99.3 kPa and a temperature of 25 °C.

   i. Calculate the molar mass of this gas. *(4 marks)*

   ii. Suggest a possible formula of this gas. *(1 mark)*

   b. Calcium nitrate $Ca(NO_3)_2$ decomposes on heating as shown below.

   $$Ca(NO_3)_2(s) \rightarrow CaO(s) + 2NO_2(g) + \tfrac{1}{2}O_2(g)$$

   A student heats and completely decomposes 0.509 g of $Ca(NO_3)_2$.

   Calculate the total volume of gas that the student would expect to collect, in cm³, measured at room temperature and pressure. *(3 marks)*

6. A student completely reacts 10.0 g of CuO with an excess of 1.75 mol dm⁻³ hydrochloric acid.

   The reaction forms aqueous copper(II) chloride and water.

   a. Write an equation, with state symbols, for the reaction of CuO with HCl. *(2 marks)*

   b. Calculate the minimum volume of hydrochloric acid needed to completely react with the CuO. *(3 marks)*

4.1 Acids, bases, and neutralisation
Specification reference: 2.1.4

Acids

Acids are substances that release H^+ ions when they are dissolved in water.

An H^+ ion is a proton and an acid is referred to as a proton donor.

Examples of common acids are shown in Table 1.

Strong acids

Hydrochloric acid, sulfuric acid, and nitric acid are strong acids.

Strong acids fully dissociate when dissolved in water:

$$HCl(g) + aq \rightarrow H^+(aq) + Cl^-(aq)$$

Weak acids

Ethanoic acid is a weak acid.

Weak acids partially dissociate when dissolved in water:

$$CH_3COOH(aq) \rightleftharpoons CH_3COO^-(aq) + H^+(aq)$$

This means that a strong acid produces more H^+ ions and has a lower pH than a weak acid with the same concentration.

▼ **Table 1** *Common acids*

Acid	Formula
hydrochloric acid	HCl
sulfuric acid	H_2SO_4
nitric acid	HNO_3
ethanoic acid	CH_3COOH

Bases and alkalis

Bases

Bases are substances that accept H^+ ions and are referred to as proton acceptors.

There are several types of base:

- metal oxides
- metal hydroxides
- metal carbonates
- alkalis.

Examples of different types of base are shown in Table 2.

▼ **Table 2** *Examples of different types of base*

Type of base	Example
metal oxides	MgO
metal hydroxides	$Ca(OH)_2$
metal carbonates	$CaCO_3$
alkali	KOH

Alkalis

Alkalis are bases that dissolve in water releasing OH^- ions, e.g.:

$$NaOH(s) + aq \rightarrow Na^+(aq) + OH^-(aq)$$

Examples of common alkalis are shown in Table 3.

▼ **Table 3** *Common alkalis*

Alkali	Formula
sodium hydroxide	NaOH
potassium hydroxide	KOH
ammonia	NH_3

Neutralisation

Neutralisation is a reaction between an acid and a base to produce a salt.
A salt is formed when the H^+ ion of an acid is replaced by a metal ion or by an ammonium ion, NH_4^+.

You must be able to write equations for neutralisation reactions of an acid with different bases: carbonates, metal oxides, and alkalis.

> **Revision tip**
> Make sure you can remember the formulae of these common acids, bases, and alkalis.

Neutralisation of an acid by a carbonate

$$\text{acid + carbonate} \rightarrow \text{salt + water + carbon dioxide}$$

$$2HCl(aq) + CaCO_3(s) \rightarrow CaCl_2(aq) + H_2O(l) + CO_2(g)$$

$$H_2SO_4(aq) + Na_2CO_3(aq) \rightarrow Na_2SO_4(aq) + H_2O(l) + CO_2(g)$$

Acids and redox

Neutralisation of an acid by a metal oxide

acid + metal oxide → salt + water

$2HCl(aq) + MgO(s) → MgCl_2(aq) + H_2O(l)$

Neutralisation of an acid by an alkali

Acid + alkali → salt + water

$HNO_3(aq) + NaOH(aq) → NaNO_3(aq) + H_2O(l)$

$H_2SO_4(aq) + 2KOH(aq) → K_2SO_4(aq) + 2H_2O(l)$

H^+ from the acid accepts OH^- from the alkali to form H_2O:

$H^+(aq) + OH^-(aq) → H_2O(l)$

Reactions with ammonia

Ammonia, NH_3, reacts with acids, accepting H^+ to form ammonium salts containing the ammonium ion, NH_4^+.

hydrochloric acid + ammonia solution → ammonium chloride

$HCl(aq) + NH_3(aq) → NH_4Cl(aq)$

sulfuric acid + ammonia solution → ammonium sulfate

$H_2SO_4(aq) + 2NH_3(aq) → (NH_4)_2SO_4(aq)$

> **Revision tip**
> In neutralisation reactions between an acid and an alkali, H^+ ions react with OH^- ions to form H_2O.

Summary questions

1. Define the term 'acid'. *(1 mark)*
2. Write the formula of:
   a. nitric acid *(1 mark)*
   b. hydrochloric acid *(1 mark)*
   c. potassium hydroxide. *(1 mark)*

3. Explain the difference between a weak acid and a strong acid. *(2 marks)*

4. Write balanced equations for the reaction between:
   a. potassium hydroxide and hydrochloric acid *(2 marks)*
   b. sodium hydroxide and sulfuric acid *(2 marks)*
   c. copper(II) oxide and nitric acid *(2 marks)*
   d. ethanoic acid and calcium carbonate. *(2 marks)*

4.2 Acid–base titrations

Specification reference: 2.1.3, 2.1.4

Acid–base titrations

Chemists use titrations to find unknown information, such as concentration, about a substance in solution.

In an acid–base titration:

- a solution of known concentration (a standard solution) is reacted with a solution of unknown concentration
- very precise apparatus (e.g. burettes and pipettes) is used to measure the volumes of solutions
- an indicator is used to show the end point, when exact neutralisation has taken place.

Preparation of a standard solution

A standard solution is a solution that has a known concentration.

For example: a $0.100 \, mol \, dm^{-3}$ solution of NaOH contains 0.100 mol of NaOH in each $1.00 \, dm^3$ of solution.

> **Synoptic link**
>
> For details of concentrations of solutions, See Topic 3.3, Moles and volumes.

> **Worked example: Preparing a standard solution**
>
> $250.0 \, cm^3$ of a $0.100 \, mol \, dm^{-3}$ solution of NaOH is to be prepared.
>
> **Step 1:** First work out the mass of NaOH that needs to be weighed out.
>
> In a $250.0 \, cm^3$ solution:
>
> amount of NaOH $n = c \times \dfrac{V}{1000} = 0.100 \times \dfrac{250}{1000} = 0.0250 \, mol$
>
> mass NaOH $m = n \times M = 0.0250 \times 40.0 = 1.00 \, g$
>
> **Step 2:** Prepare the solution in a volumetric flask.
>
> - Weigh out 1.00 g of sodium hydroxide and add to a beaker.
> - Dissolve the NaOH, with stirring, in distilled (or deionised) water.
> - Pour the solution into a $250.0 \, cm^3$ volumetric flask.
> - Rinse the beaker with distilled water and wash the rinsings into the volumetric flask.
> - Add distilled water to the volumetric flask until the bottom of the meniscus is exactly on the graduation line.
> - Place the stopper in the volumetric flask and invert the flask several times to thoroughly mix the solution.

> **Revision tip**
>
> Make sure the bottom of the meniscus is viewed at eye level.

Carrying out an acid–base titration

The procedure below describes how to carry out a titration to find the volume of sulfuric acid, H_2SO_4, that reacts with $25.0 \, cm^3$ of a solution of sodium hydroxide, NaOH.

1. Use a pipette to add $25.0 \, cm^3$ of NaOH(aq) into a conical flask.
2. Place the flask on a white tile and add several drops of indicator.
3. Add the sulfuric acid to the burette and record the initial burette reading to the nearest $0.05 \, cm^3$.
4. Add the sulfuric acid from the burette to the flask, swirling the flask throughout to mix the contents.

Acids and redox

5 The colour of the indicator changes at the end point of the titration. When the colour changes, stop adding the acid and record the final burette reading to the nearest $0.05\,cm^3$.

6 Calculate the volume of acid added by subtracting the initial burette reading from the final burette reading. This is called the **trial titre** and gives you a rough idea of how much acid is required for neutralisation.

7 Repeat the whole procedure, adding the sulfuric acid dropwise as the end point is approached, to obtain an accurate titre.

8 Repeat the titration until you have obtained two titres within $0.10\,cm^3$. These results are said to be concordant.

9 Average the accurate titres that are concordant to determine the mean titre. You will use the mean titre to analyse the results.

Acid–base titration calculations

Reactions take place in molar proportions which can be determined from a balanced equation. For example, NaOH reacts with HCl in a 1:1 molar ratio but NaOH reacts with H_2SO_4 in a 2:1 molar ratio:

$$NaOH(aq) + HCl(aq) \rightarrow NaCl(aq) + H_2O(l)$$
$$1\,mol \quad\quad 1\,mol$$

$$2NaOH(aq) + H_2SO_4(aq) \rightarrow Na_2SO_4(aq) + 2H_2O(l)$$
$$2\,mol \quad\quad 1\,mol$$

When titration results are analysed, you need to take into account the molar ratio. The analysis of titration results follows a set pattern.

 Worked example: Calculating an unknown concentration from titration results

$25.00\,cm^3$ of sodium hydroxide solution is exactly neutralised by $21.40\,cm^3$ of $0.0500\,mol\,dm^{-3}$ sulfuric acid. What is the concentration of the sodium hydroxide solution?

Step 1: Write a balanced equation.

$$2NaOH(aq) + H_2SO_4(aq) \rightarrow Na_2SO_4(aq) + 2H_2O(l)$$

Step 2: Add the data given in the question under the reactants in the equation.

$$2NaOH(aq) + H_2SO_4(aq) \rightarrow Na_2SO_4(aq) + 2H_2O(l)$$
$$25.00\,cm^3 \quad\quad 21.40\,cm^3$$
$$? \quad\quad\quad\quad 0.0500\,mol\,dm^{-3}$$

Step 3: Calculate the amount, in mol, of H_2SO_4

$$\text{Amount } n = c \times \frac{V}{1000} = 0.0500 \times \frac{21.40}{1000} = 0.00107\,mol$$

Step 4: Use the balanced equation to determine the amount, in mol, of NaOH:

2 mol NaOH reacts with 1 mol H_2SO_4

∴ amount, in mol, of NaOH = $2 \times 0.00107 = 0.00214\,mol$

Step 5: Calculate the unknown concentration of NaOH.

$$\text{Amount } n = c \times \frac{V}{1000}$$

∴ Concentration $c = n \times \frac{1000}{V} = 0.00214 \times \frac{1000}{25.0} = 0.0856\,mol\,dm^{-3}$

Summary questions

1 Calculate the following. Remember to show your working.
   a The concentration of a $1.00\,dm^3$ solution containing $6.00\,g$ of NaOH. *(1 mark)*
   b The concentration of a $250\,cm^3$ solution containing $3.20\,g$ of CH_3COOH. *(2 marks)*
   c The amount in mol of HNO_3 in $21.7\,cm^3$ of $0.200\,mol\,dm^{-3}$ $HNO_3(aq)$. *(1 mark)*
   d The volume of $2.00\,mol\,dm^{-3}$ KOH that contains $0.250\,mol$ of KOH. *(1 mark)*

2 $25.0\,cm^3$ of NaOH(aq) is neutralised by $27.60\,cm^3$ of $0.150\,mol\,dm^{-3}$ $HNO_3(aq)$. Calculate the concentration of NaOH(aq), to **three** significant figures. *(3 marks)*

3 $25.0\,cm^3$ of $2.44 \times 10^{-2}\,mol\,dm^{-3}$ HCl is exactly neutralised by $24.20\,cm^3$ of $Ca(OH)_2$ solution.
   a Write the equation for the reaction. *(1 mark)*
   b Calculate the concentration of the $Ca(OH)_2$ solution. *(3 marks)*

18

4.3 Redox
Specification reference: 2.1.5

Assigning oxidation numbers

An oxidation number has a + sign or a − sign followed by a number.

Oxidation numbers are assigned to each atom in an element, ion, or compound using a set of rules.

Oxidation number rules
Elements
In a pure element, the oxidation number is zero.

Ions
In monatomic ions, the oxidation number is equal to the ionic charge:

- Halide ions (e.g. Br⁻) have an oxidation number of −1.
- Group 1 metal ions (e.g. Na⁺) have an oxidation number of +1.
- Group 2 metal ions (e.g. Ca^{2+}) have an oxidation number of +2.

Compounds
In compounds,

- Fluorine always has an oxidation number of −1.
- Hydrogen has an oxidation number of +1.
- Oxygen has an oxidation number of −2.

Common exceptions
Hydrogen:

- in a metal hydride (e.g. NaH), hydrogen has an oxidation number of −1.

Oxygen:

- when bonded to fluorine (e.g. F_2O), oxygen has an oxidation number of +2.
- in a peroxide (e.g. H_2O_2), oxygen has an oxidation number of −1.

Oxidation numbers in compounds and polyatomic ions
In a neutral compound,

- sum of oxidation numbers = zero:
  e.g. in CO_2, $+4 + (-2 \times 2) = 0$

In a polyatomic ion,

- sum of oxidation numbers = overall charge on the ion:
  e.g. in CO_3^{2-}, $+4 + (-2 \times 3) = -2$

Elements with variable oxidation numbers

Roman numerals are used to show the oxidation number of elements that can have different oxidation numbers in different compounds or ions.

In iron(II) oxide, FeO, oxidation number of Fe is +2.

In iron(III) oxide, Fe_2O_3, oxidation number of Fe is +3.

In sodium chlorate(I), NaClO, oxidation number of Cl is +1.

In sodium chlorate(III), $NaClO_2$, oxidation number of Cl is +3.

> **Revision tip**
> **Common error 1**
> Confusing oxidation numbers in a pure element and in an ion:
> - In the pure element, oxidation number of Ca is zero.
> - In an ionic compound, e.g. $CaCl_2$, oxidation number of Ca is +2.
>
> **Common error 2**
> Not assigning oxidation numbers to each atom:
> - In H_2O, there are two H atoms; each H atom has an oxidation number of +1.
> - The oxidation number of H is *not* +2!

> **Revision tip**
> In an oxidation number, the sign comes before the number.
>
> In an ionic charge, the number comes before the sign.

> **Synoptic link**
> You first came across the use of Roman numerals in formulae in Topic 2.3, Compounds, formulae, and equations.

> **Revision tip**
> In these examples, the formula is given but you can work out the formula from using the name and the Roman numeral. Try working out the formula for sodium chlorate(V) and sodium chlorate(VII).

Acids and redox

Oxidation and reduction

Oxidation and reduction occur in many chemical reactions. Oxidation and reduction always occur together and the reaction is called a 'redox reaction'.

Oxidation and reduction in terms of electron transfer

- Oxidation is the loss of electrons.
- Reduction is the gain of electrons.

Total number of electrons lost during oxidation
= total number of electrons gained during reduction.

Oxidation and reduction in terms of oxidation number

- Oxidation is an increase in oxidation number.
- Reduction is a decrease in oxidation number.

Total increase in oxidation number during oxidation
= total decrease in oxidation number during reduction.

Redox reactions of metals and acids

When zinc is added to hydrochloric acid, a redox reaction takes place:

$$Zn + 2HCl \rightarrow ZnCl_2 + H_2$$

$$0 \quad \rightarrow +2 \quad \text{oxidation}$$
$$+1 \rightarrow \quad 0 \quad \text{reduction}$$

- The oxidation number of Zn has increased by 2 and Zn has been oxidised.
- The oxidation number of each H has decreased by 1 and H has been reduced.

Two H atoms have been reduced from +1 to 0 and so the total decrease is $2 \times 1 = 2$.

The total increase of 2 in oxidation number is balanced by the total decrease of 2.

Identifying redox reactions

Magnesium reacts with copper(II) sulfate solution in a redox reaction:

$$Mg + CuSO_4 \rightarrow MgSO_4 + Cu$$

$$0 \quad \rightarrow +2 \quad \text{oxidation}$$
$$+2 \rightarrow \quad 0 \quad \text{reduction}$$

- The oxidation number of Mg has increased by 2 and magnesium has been oxidised.
- The oxidation number of Cu has decreased by 2 and Cu has been reduced.

The total increase of 2 in oxidation number is balanced by the total decrease of 2.

Half equations

You can show the electron transfer in a redox reaction by splitting the overall equation into two half equations.

In the reaction of magnesium and copper(II) sulfate solution:

$$Mg \rightarrow Mg^{2+} + 2e^- \quad \text{oxidation: loss of 2 electrons}$$
$$Cu^{2+} + 2e^- \rightarrow Cu \quad \text{reduction: gain of 2 electrons}$$

Total number of electrons lost = total number of electrons gained.

Key term

Redox reaction: A reaction involving reduction and oxidation.

Revision tip

In oxidation, electrons are lost and oxidation number increases.

In reduction, electrons are gained and oxidation number decreases.

Revision tip

The oxidation and reduction processes must balance in terms of electrons transferred and changes in oxidation number.

Synoptic link

You will learn much more about redox reactions in Year 2 of the A Level course.

Summary questions

1 Define 'oxidation' and 'reduction', in terms of electron transfer and oxidation number.
   (2 marks)

2 What is the oxidation number in the following?
   a oxygen in O_2 *(1 mark)*
   b copper in Cu_2O *(1 mark)*
   c sulfur in SO_4^{2-} *(1 mark)*

3 In the reactions below, use oxidation numbers to show which element has been oxidised and which has been reduced.
   a $4Na + O_2 \rightarrow 2Na_2O$ *(2 marks)*
   b $2Al + 3H_2SO_4 \rightarrow Al_2(SO_4)_3 + 3H_2$ *(2 marks)*
   c $MnO_2 + 4HCl \rightarrow MnCl_2 + Cl_2 + 2H_2O$ *(2 marks)*

Chapter 4 Practice questions

1. Which reaction of sulfuric acid does **not** represent a neutralisation reaction?
    A $2NaOH + H_2SO_4 \rightarrow Na_2SO_4 + 2H_2O$
    B $NH_3 + H_2SO_4 \rightarrow (NH_4)_2SO_4$
    C $MgCO_3 + H_2SO_4 \rightarrow MgSO_4 + H_2O + CO_2$
    D $Fe + H_2SO_4 \rightarrow FeSO_4 + H_2$ *(1 mark)*

2. Which equation shows sulfur being oxidised.
    A $H_2SO_4 + 8HI \rightarrow 4H_2O + H_2S + 4I_2$
    B $H_2SO_4 + 2HBr \rightarrow 2H_2O + SO_2 + Br_2$
    C $2H_2O + 2SO_2 + O_2 \rightarrow 2H_2SO_4$
    D $H_2SO_3 + 2NaOH \rightarrow Na_2SO_3 + 2H_2O$ *(1 mark)*

3. What is the name of the salt $NaClO_3$?
    A sodium chloride(III)   B sodium chloride(V)
    C sodium chlorate(III)   D sodium chlorate(V) *(1 mark)*

4. Chlorine reacts with sodium iodide as below.
    $$Cl_2 + 2NaBr \rightarrow 2NaCl + Br_2$$
    Which statement about this reaction is correct? *(1 mark)*
    A Bromide ions gain electrons.
    B Chlorine has been reduced.
    C The oxidation number of bromine decreases.
    D The oxidation number of chlorine increases.

5. A student is provided with a sample of nitric acid of unknown concentration. The student titrates the nitric acid with $0.1500\,mol\,dm^{-3}$ barium hydroxide, $Ba(OH)_2$. The equation is shown below.
    $$Ba(OH)_2(aq) + 2HNO_3(aq) \rightarrow Ba(NO_3)_2(aq) + 2H_2O(l)$$
    $25.00\,cm^3$ of HNO_3 is neutralised by $30.00\,cm^3$ of $0.1500\,mol\,dm^{-3}$ $Ba(OH)_2$.
    a Calculate the concentration, in $mol\,dm^{-3}$, of the nitric acid. *(3 marks)*
    b Barium hydroxide is an alkali.
       What is meant by an *alkali*? *(1 mark)*
    c Write the ionic equation for this reaction. *(1 mark)*
    d Nitric acid can also be neutralised by aqueous sodium carbonate.
       Write the equation for this reaction. *(1 mark)*

6. Aluminum chloride, $AlCl_3$, is a salt of hydrochloric acid.
    a Aluminum chloride can be prepared by reacting aluminum with hydrochloric acid. This is a redox reaction. The unbalanced equation is shown below.
       $$Al(s) + HCl(aq) \rightarrow AlCl_3(aq) + H_2(g)$$
       i Balance this equation. *(1 mark)*
       ii Use oxidation numbers to identify which element has been oxidised and which element has been reduced. Explain your answer. *(2 marks)*
       iii State **two** observations for this reaction. *(2 marks)*
    b Aluminum sulfate can also be prepared by reacting aluminum oxide, Al_2O_3, with sulfuric acid.
       i Write an equation, including state symbols, for this reaction. *(2 marks)*
       ii What type of reaction has taken place? *(1 mark)*

5.1 Electron structure

Specification reference: 2.2.1

▼ **Table 1** *The number of electrons that fill the first four shells*

Shell number, n	Number of electrons
1	2
2	8
3	18
4	32

Key term

Orbital: A region around the nucleus that can hold up to two electrons, with opposite spins.

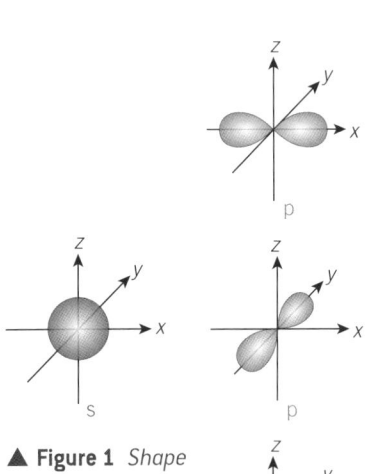

▲ **Figure 1** *Shape of an s-orbital*

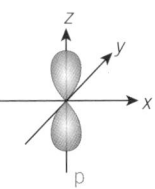

▲ **Figure 2** *Shape of p-orbitals*

Revision tip

You need to know the shapes of s- and p-orbitals.

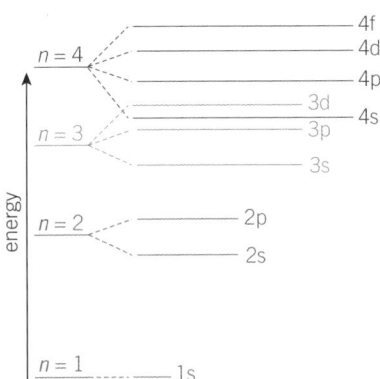

▲ **Figure 3** *Energy levels of sub-shells*

Shells (Energy levels)

Electrons are in constant motion around the nucleus of an atom.

- Electrons are found in energy levels or shells.
- Shells are made up of sub-shells.
- Sub-shells are made up of orbitals.

Table 1 shows the number of electrons that fill the first four shells.

Atomic orbitals

An atomic orbital is a region around the nucleus that can hold up to two electrons, with opposite spins.

- There are different types of orbital, s, p, d, and f, with different shapes. Figure 1 and Figure 2 show the shapes of s and p orbitals.
- The shape of an orbital shows where an electron is most likely to be located.

Sub-shells

- Within a shell, the different types of orbital are grouped together as sub-shells.
- Different types of sub-shell contain different numbers of orbitals.

Table 2 summarises the orbitals, sub-shells, and orbitals in the first four shells.

▼ **Table 2** *Sub-shells and orbitals for the first four shells*

Shell	1	2		3			4			
Electrons in shell	2	8		18			32			
Sub-shell	1s	2s	2p	3s	3p	3d	4s	4p	4d	4f
Number of orbitals	1	1	3	1	3	5	1	3	5	7
Electrons in sub-shell	2	2	6	2	6	10	2	6	10	14
Electron configuration	$1s^2$	$2s^2$	$2p^6$	$3s^2$	$3p^6$	$3d^{10}$	$4s^2$	$4p^6$	$4d^{10}$	$4f^{14}$

Electron configurations

We use three principles for working out electron configurations.

1. Electrons fill orbitals in order of increasing energy.
   Figure 3 shows the order of increasing energy of the sub-shells in the first four shells.
   Note that the 4s sub-shell has a lower energy than the 3d sub-shell, so 4s is filled before 3d.

2. An orbital can contain a maximum of two electrons with opposite spins, shown as an up arrow and a down arrow.

3. For orbitals of the same energy, orbitals are occupied singly before pairing.

Electron configurations of atoms

Using these principles, you can write electron configurations of atoms and show electrons in boxes diagrams for the direction of the electron spins. Table 3 shows examples.

▼ **Table 3** *Electron configurations of a selection of atoms*

Element	Electron configuration	'Electrons in boxes' diagrams				
		1s	2s	2p	3s	3p
H	$1s^1$	↑				
He	$1s^2$	↑↓				
Li	$1s^2 2s^1$	↑↓	↑			
C	$1s^2 2s^2 2p^2$	↑↓	↑↓	↑ ↑		
O	$1s^2 2s^2 2p^4$	↑↓	↑↓	↑↓ ↑ ↑		
P	$1s^2 2s^2 2p^6 3s^2 3p^3$	↑↓	↑↓	↑↓ ↑↓ ↑↓	↑↓	↑ ↑ ↑

Blocks of the periodic table

- s-block elements, e.g. Na, have their highest energy electrons in an s sub-shell.
- p-block elements, e.g. N, have their highest energy electrons in a p sub-shell.
- d-block elements, e.g. V, have their highest energy electrons in a d sub-shell.

Electron configurations of ions

Atoms gain or lose electrons to form ions. Metal atoms lose electrons to form positive ions. Non-metal atoms gain electrons to form negative ions.

▼ **Table 4** *The electron configurations of a selection of atoms and ions*

Element	Lithium Li	Aluminium Al	Oxygen O	Fluorine F
Group	1 (1)	13 (3)	16 (6)	17 (7)
Electron configuration of atom	$1s^2 2s^1$	$1s^2 2s^2 2p^6 3s^2 3p^1$	$1s^2 2s^2 2p^4$	$1s^2 2s^2 2p^5$
Ion	Li^+	Al^{3+}	O^{2-}	F^-
Electron configuration of ion	$1s^2$	$1s^2 2s^2 2p^6$	$1s^2 2s^2 2p^6$	$1s^2 2s^2 2p^6$

Ions of d-block elements

When d-block elements form ions, they lose their highest energy electrons.

When the sub-shells are empty the 4s sub-shell is lower in energy than the 3d sub-shell, so the 4s orbital is occupied before the 3d orbitals.

Once the orbitals are occupied, the 4s electrons are higher in energy than the 3d electrons. The 4s electrons are lost before the 3d electrons.

Examples

Ti atom: $1s^2 2s^2 2p^6 3s^2 3p^6 3d^2 4s^2$ → Ti^{2+} ion: $1s^2 2s^2 2p^6 3s^2 3p^6 3d^2$

Mn atom: $1s^2 2s^2 2p^6 3s^2 3p^6 3d^5 4s^2$ → Mn^{2+} ion: $1s^2 2s^2 2p^6 3s^2 3p^6 3d^5$

Electrons and bonding

Revision tip
Notice that orbitals in the same sub-shell are occupied singly before pairing.

In Table 3, see the orbitals for the p sub-shells of C, O, and P.

Synoptic link
You will learn more about the link between electron configuration and the periodic table in Topic 7.1, The periodic table.

Revision tip
The electron configurations of ions often have full shells.

Revision tip
The 4s sub-shell is filled and emptied before the 3d sub-shell.

Summary questions

1 Describe the shape of an s- and a p-orbital. *(1 mark)*

2 Write full electron configurations for the following atoms:
  a Na b S c Cl
  d Be e Fe f Ni *(6 marks)*

3 Write full electron configurations for the following ions:
  a Mg^{2+} b S^{2-} c O^{2-}
  d Na^+ e Sc^{3+} f Zn^{2+} *(6 marks)*

5.2 Ionic bonding and structure

Specification reference: 2.2.2

Formation of ions

All chemical bonds are forces of attraction between oppositely charged particles.

- Metal atoms lose electrons forming **positive ions** (cations), e.g. a lithium atom loses 1 electron to form a lithium ion:

$$Li \rightarrow Li^+ + e^-$$
$$1s^2 2s^1 \rightarrow 1s^2$$

A Li^+ ion has the same electron configuration as a helium atom.

- Non-metal atoms gain electrons forming **negative ions** (anions), e.g. a fluorine atom gains 1 electron to form a fluoride ion.

$$F + e^- \rightarrow F^-$$
$$1s^2 2s^2 2p^5 \rightarrow 1s^2 2s^2 2p^6$$

A F^- ion has the same electron configuration as a neon atom.

The transfer of electrons from metal to non-metal often forms ions that have a full noble gas electron configuration. The number of electrons transferred may be different depending on the group in the periodic table.

Ionic bonding

Ionic bonds occur when a metal and a non-metal react to form a compound.

An ionic bond is the electrostatic attraction between positive and negative ions.

The ions in ionic compounds can be shown in 'dot-and-cross' diagrams. Figure 1 shows three examples: LiF, MgO and Al_2S_3.

$[Li]^+ \; [\overset{xx}{\underset{xx}{×F×}}]^-$	$[Mg]^{2+} \; [\overset{••}{\underset{••}{×O×}}]^{2-}$	$2[Al]^{3+} \; 3[\overset{••}{\underset{••}{×S×}}]^{2-}$
LiF	MgO	Al_2S_3

▲ **Figure 1** *'Dot-and-cross' diagrams*

Giant ionic lattices

The ions in an ionic compound form a repeating three-dimensional structure, called a giant ionic lattice.

- In an ionic compound, each ion is surrounded by ions of opposite charge.
- For example, in the sodium chloride lattice, each Cl^- ion is surrounded by six Na^+ ions and vice versa. Oppositely charged ions are attracted to each other in all directions. See Figure 2 and Figure 3.
- This attraction is repeated throughout the lattice structure.

> **Synoptic link**
>
> For more details of writing the formula for ionic compounds, see Topic 2.3 Compounds, formulae, and equations.

> **Key term**
>
> **Ionic bond:** An electrostatic attraction between ions of opposite charge.

> **Revision tip**
>
> When drawing 'dot-and-cross' diagrams, you only need to draw the outer shell electrons. Notice that the ionic charge is shown outside of the square brackets.

> **Revision tip**
>
> Empirical formulae are used for compounds with giant ionic lattices. See Topic 2.3, Compounds, formula, and equations, and Topic 3.2, Determination of formulae.

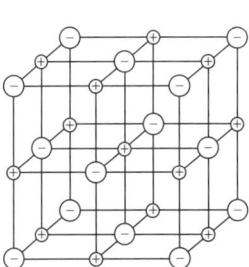

▲ **Figure 2** *Sodium chloride structure*

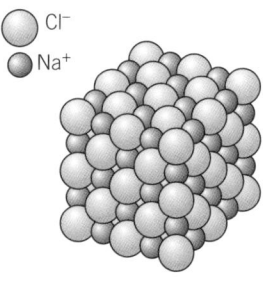

▲ **Figure 3** *A space-filling diagram showing the giant ionic lattice structure of sodium chloride*

Properties of ionic compounds

Melting and boiling points

A large amount of energy would be required to break all the strong ionic bonds in the giant ionic lattice. As a result, ionic compounds generally have high melting and boiling points, and are solids at room temperature.

The strength of ionic bonding depends on the size and charge of the ions involved:

- Smaller ions form stronger ionic bonds.
- More highly-charged ions form stronger ionic bonds.

Solubility

Ionic compounds are soluble in polar solvents like water, but insoluble in non-polar solvent like hydrocarbons.

Electrical conductivity

Ionic compounds do not conduct electricity when solid because the ions are in fixed positions in the giant ionic lattice.

When molten or dissolved in water, ionic compounds do conduct electricity. The giant ionic lattice has broken down and the ions can now move.

> **Synoptic link**
>
> You will find out more about polarity in Topic 6.2, Electronegativity and polarity.

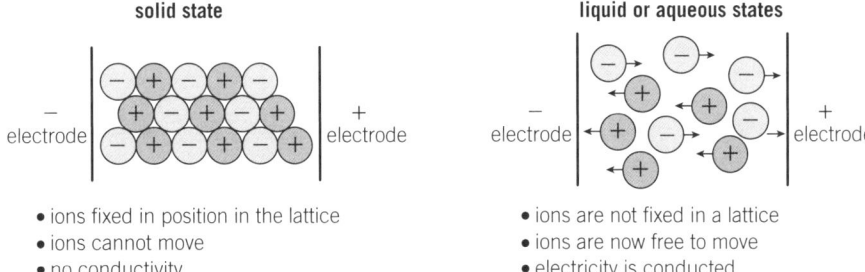

- ions fixed in position in the lattice
- ions cannot move
- no conductivity

- ions are not fixed in a lattice
- ions are now free to move
- electricity is conducted

▲ **Figure 4** *Electrical conductivity of an ionic compound in solid, liquid, and aqueous states*

Summary questions

1. What is meant by the term 'ionic bond'? *(1 mark)*

2. Why does sodium chloride conduct electricity when molten but not when solid? *(2 marks)*

3. Magnesium phosphide is an ionic compound.
   a State the electron configuration of the ions present in magnesium phosphide
   b Predict the formula of magnesium phosphide. *(2 marks)*

4. Draw 'dot-and-cross' diagrams, showing outer electrons only, for the following compounds.
   a calcium chloride
   b potassium nitride
   c iron(III) sulfide. *(6 marks)*

5.3 Covalent bonding

Specification reference: 2.2.2

Key terms

Covalent bond: A shared pair of electrons.

Molecule: a small unit of two or more atoms covalently bonded together.

The covalent bond

Covalent bonds exist between non-metal atoms by sharing pairs of electrons. The atoms often then obtain a full noble gas electron configuration.

- A covalent bond is a shared pair of electrons.
- In a covalent bond, the atoms are held together by strong electrostatic attraction between the shared pair of electrons and the nuclei of the bonded atoms.

Many covalent compounds exist as simple molecules: a small unit of two or more atoms covalently bonded together.

Covalent bonds can be represented in 'dot-and-cross' diagrams. A covalent bond can also be represented by a straight line between the atoms. If all bonds are shown with lines, the diagram is called a **displayed formula**.

Synoptic link

For more details of displayed formulae, see Topic 11.3, Representing the formulae of organic compounds.

H : H	H—H	H : C : H (with H above and below)	H—C—H (with H above and below)
H_2		CH_4	

▲ Figure 1 *Covalent bonds in hydrogen and methane molecules*

Multiple covalent bonds

Non-metal atoms can form double or triple covalent bonds by sharing more than one pair of electrons. These are represented in displayed formulae by multiple lines between the atoms. Double and triple bonds are stronger than single bonds.

O::C::O	O=C=O	:N:::N:	N≡N
CO_2		N_2	

▲ Figure 2 *Double and triple covalent bonds in carbon dioxide and nitrogen molecules*

Key terms

Double covalent bond: Two shared pairs of electrons.

Triple covalent bond: Three shared pairs of electrons.

Dative covalent (coordinate) bonds

Lone pairs

Lone pairs of electrons are paired electrons that are not involved in bonding.

In an ammonia, NH_3, molecule, the nitrogen atom has three covalent bonds to the three hydrogen atoms and one lone pair of electrons.

In a water, H_2O, molecule, the oxygen atom has two covalent bonds to the two hydrogen atoms and two lone pairs of electrons.

Synoptic link

You will learn more about the nature of a double bond when you study organic chemistry in Topic 13.1, The properties of the alkenes.

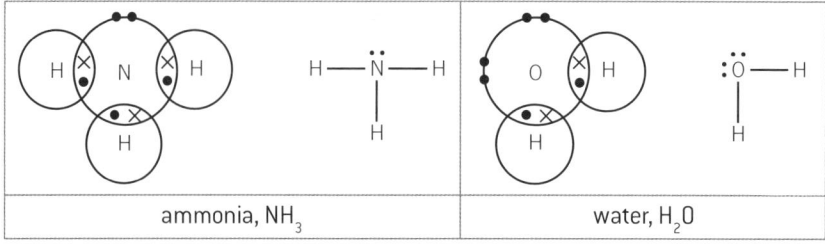

▲ Figure 3 *Lone pairs of electrons in NH_3 and H_2O*

Electrons and bonding

Bonding with lone pairs

Lone pairs of electrons are able to form dative covalent bonds with atoms that have vacant orbitals.

- A dative covalent bond is a shared pair of electrons in which both bonded electrons are contributed by one of the atoms in the bond.
- Dative covalent bonds are often shown in displayed formulae by an arrow.

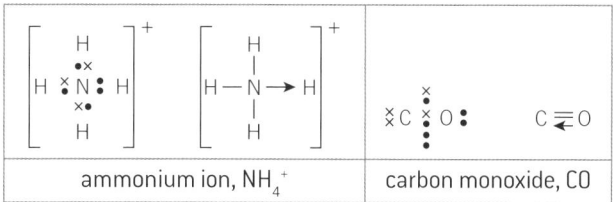

▲ Figure 4 *Dative covalent bonds in NH_4^+ and CO*

Key terms

Lone pair: A pair of electrons that is not involved in bonding.

Dative covalent (or co-ordinate) bond: A shared pair of electrons in which both bonded electrons are contributed by one of the atoms in the bond.

Molecules without an octet of electrons

Less than eight electrons

Boron has only three electrons in its outer shell, and forms three covalent bonds. In boron trifluoride, BF_3, boron has six electrons in its outer shell (see Figure 5).

More than eight electrons

Elements in Period 3 of the periodic table are able to have more than eight electrons in their outer shell. This is because they have access to the d sub-shell.

For example, in sulfur dioxide, SO_2, sulfur has 10 electrons in its outer shell by pairing four of its six electrons (see Figure 5). Note that oxygen cannot expand beyond eight electrons as it is a Period 2 element.

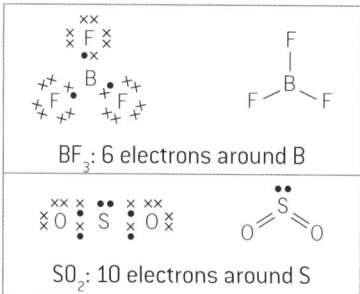

▲ Figure 5 *Molecules without an octet of electrons*

Average bond enthalpy

Average bond enthalpy is a measurement of covalent bond strength. The larger the value of the average bond enthalpy, the stronger the covalent bond.

Synoptic link

Average bond enthalpy is covered in detail in Topic 9.3, Bond enthalpies. The role of bond enthalpies in rates of reaction is studied in Topic 15.1, The chemistry of the haloalkanes.

Summary questions

1 a What is a covalent bond? *(1 mark)*
   b What is a dative covalent bond? *(1 mark)*

2 Draw 'dot-and-cross' diagrams for the following.
   a NF_3 *(1 mark)*
   b $COCl_2$ (C is the central atom) *(1 mark)*
   c HCN (C is the central atom). *(1 mark)*

3 Draw 'dot-and-cross' diagrams for the following. Show any 'extra' electrons with a triangle.
   a H_2F^+ *(2 marks)*
   b NO_2^- (N is central atom) *(2 marks)*
   c CO_3^{2-} (C is the central atom). *(2 marks)*

Chapter 5 Practice questions

1 How many sub-shells are full in a calcium atom?
   A 3 **B** 4
   C 5 **D** 6 *(1 mark)*

2 How many electrons are unpaired in a phosphorus atom?
   A 1 **B** 2
   C 3 **D** 4 *(1 mark)*

3 What is the electron configuration of a zinc ion, Zn^{2+}?
   A $1s^2 2s^2 2p^6 3s^2 3p^6 3d^{10}$
   B $1s^2 2s^2 2p^6 3s^2 3p^6 3d^8 4s^2$
   C $1s^2 2s^2 2p^6 3s^2 3p^6 3d^{10} 4s^2$
   D $1s^2 2s^2 2p^6 3s^2 3p^6 3d^8$ *(1 mark)*

4 This question looks at electron structure in atoms and ions.

   a Complete the table for the following regions when they are full.

	Number of orbitals	Number of electrons
The 2p sub-shell		
The 3rd shell		

   (2 marks)

   b Write the electron configuration for the following.
   i A bromide ion *(1 mark)*
   ii A Co^{3+} ion *(1 mark)*

   c An ionic compound **A** forms between two elements that have atoms with the electron configurations:
   $1s^2 2s^2 2p^6 3s^2 3p^3$
   $1s^2 2s^2 2p^6 3s^2 3p^6 4s^2$
   i What is the formula of compound **A**? *(1 mark)*
   ii The two ions in the compound **A** are 'isoelectronic'.
   Suggest the meaning of *isoelectronic*. Use the ions in compound **A** to help your answer. *(2 marks)*

5 Oxygen forms compounds containing ionic and covalent bonds.

   a What is meant be the following terms?
   i An ionic bond *(1 mark)*
   ii An covalent bond *(1 mark)*

   b Draw 'dot-and-cross' diagrams to show the bonding in the following compounds and ions of oxygen.
   Show outer electrons only.
   i CaO *(2 marks)*
   ii F_2O *(1 mark)*
   iii H_3O^+ *(1 mark)*

   c Sodium chloride, NaCl, has properties of a typical ionic compound. Explain the following properties.
   i NaCl has a high melting point. *(2 marks)*
   ii NaCl does not conduct electricity when solid but does conducts electricity when dissolved in water. *(2 marks)*
   iii NaCl dissolves in water. *(2 marks)*

6.1 Shapes of molecules and ions

Specification reference: 2.2.2

Valence shell electron pair repulsion theory

The shapes of molecules and ions are determined by applying valence shell electron pair repulsion theory.

- Valence shell electrons are the electrons in the outermost shell.
- The electron pairs in the valence shell repel each other as far apart as possible.
- Multiple bonds have a similar repelling effect to single bonds.
- Lone pairs of electrons repel more than bonded pairs of electrons, reducing bond angles slightly.

Shapes of molecules and ions

To work out the shape of a molecule or ion, you must first draw a 'dot-and-cross' diagram, showing electrons in the outer shell only. A 'dot-and-cross' diagram shows the number of electron pairs around the central atom and you will then be able to determine the shape.

The atoms in the molecule or ion describe the shape, after taking into account the electron-repelling effects of electron pairs.

Molecules with bonded pairs

Table 1 and Figure 1 show the shapes and bond angles of molecules with different numbers of bonded pairs or bonded regions around the central atom.

▼ **Table 1** Shapes and bond angles

Bonded pairs/regions	Example	Shape	Bond angles
2	CO_2	linear	180°
3	BF_3	trigonal planar	120°
4	CH_4	tetrahedral	109.5°
6	SF_6	octahedral	90°

Molecules with bonded pairs and lone pairs

Ammonia, NH_3, and water, H_2O

Ammonia and water molecules have four pairs of electrons around the central atom.

- In ammonia, the central N atom has 3 bonded pairs and 1 lone pair.
- In water, the central O atom has 2 bonded pairs and 2 lone pairs.

The four electron pairs repel one another but lone pairs of electrons repel more than bonded pairs.

- NH_3 molecules have a pyramidal shape with a bond angle of 107°.
- H_2O molecules have a non-linear shape with a bond angle of 104.5°.

Both bond angles are slightly less than the 109.5° bond angle in methane.

▲ **Figure 2** The shapes of ammonia and water molecules

Synoptic link

For details of 'dot-and-cross' diagrams of molecules, see Topic 5.3, Covalent bonding.

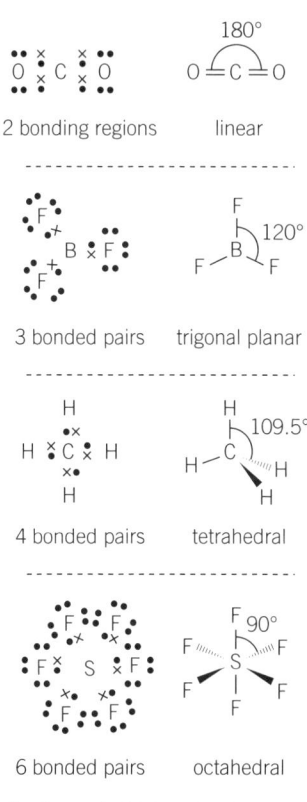

▲ **Figure 1** Molecular shapes

Revision tip

There is a lot of important information in Table 1 and Figure 1 that sums up many of the key principles for this whole topic.

Revision tip

The extra repelling effect of each lone pair reduces the bond angle by about 2.5°.

Revision tip

Methane, CH_4, has a bond angle of 109.5°.

Ammonia, NH_3, has a bond angle of 107°.

Water, H_2O, has a bond angle of 104.5°.

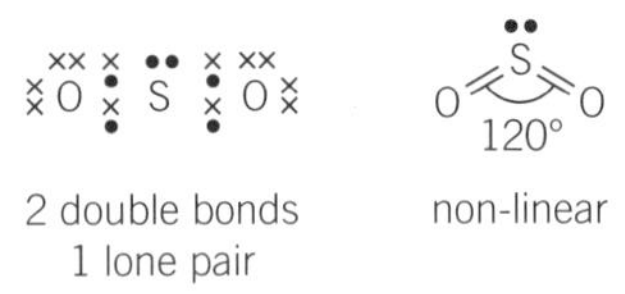

Shapes of molecules and intermolecular forces

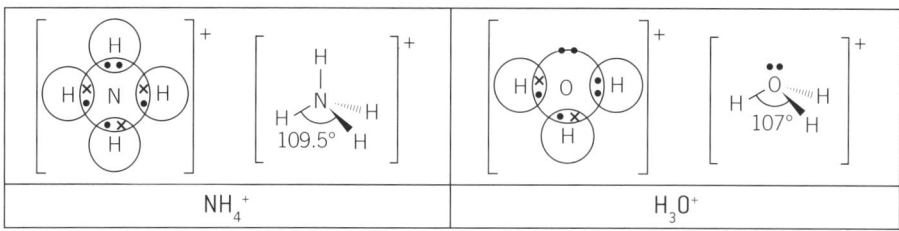

2 double bonds
1 lone pair

non-linear

▲ Figure 3 *The shape of a sulfur dioxide molecule*

Sulfur dioxide, SO$_2$

Sulfur dioxide molecules have two double bonds and one lone pair around the central sulfur atom. The double bonds have a similar repelling effect to the lone pair and the three regions of electron density repel one another as far apart as possible.

As a result, a sulfur dioxide molecule has a non-linear shape and the O=S=O bond angle is 120°.

Shapes of ions

The shapes of polyatomic ions can be worked out in the same way as for molecules.

- Remember to take account of the ionic charge when drawing the 'dot-and-cross' diagram.
- Positively charged ions have fewer electrons than the original atoms.
- Negatively charged ions have gained extra electrons, shown using triangles on a 'dot-and-cross' diagram.

Ammonium, NH$_4^+$, and hydronium, H$_3$O$^+$ ions

- In the NH$_4^+$ ion, there is a dative covalent bond between the central N atom and one of the H atoms.
- In the H$_3$O$^+$ ion, there is a dative covalent bond between the central O atom and one of the H atoms.

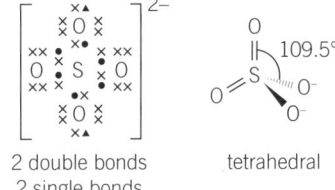

▲ Figure 4 *The shapes of NH$_4^+$ and H$_3$O$^+$ ions*

The sulfate, SO$_4^{2-}$ ion

The sulfate ion, SO$_4^{2-}$, has a complex 'dot-and-cross' diagram (Figure 5):

- The central S atom is bonded to the four O atoms by two double bonds and two single bonds.
- The two O atoms with single bonds each contain one 'extra' electron, shown by a triangle. This gives each O atom a full outer shell.
- The 2– charge on the ion results from the two extra electrons.

2 double bonds
2 single bonds

tetrahedral

▲ Figure 5 *The shape of a SO$_4^{2-}$ ion*

The SO$_4^{2-}$ ion has two double bonds and two single bonds around the central sulfur atom The double bonds have a similar repelling effect to the single bonds and the four regions of electron density repel one another as far apart as possible. As a result, the SO$_4^{2-}$ ion has a tetrahedral shape with equal bond angles of 109.5°.

Summary questions

1. How are the shapes of molecules and ions determined? *(2 marks)*

2. Draw 3D diagrams for each molecule, name the shape, and state the bond angles.
   - **a** BH$_3$
   - **b** CCl$_4$
   - **c** CS$_2$
   - **d** PH$_3$
   - **e** H$_2$S

   (15 marks)

3. Draw a 'dot-and-cross' diagram for a BF$_4^-$ ion, and state its shape and bond angle. *(3 marks)*

6.2 Electronegativity and polarity
Specification reference: 2.2.2

Electronegativity

A covalent bond is the attraction between a shared pair of electrons and the nuclei of the two bonded atoms.

- When the bonded atoms are the same, the pair of electrons in the covalent bond is equally shared, e.g. H–H, Cl–Cl.
- When the bonded atoms are different, the sharing of electrons may be unequal, e.g. in H–F, the F atom attracts the shared pair more than the H atom.

Electronegativity is the ability of an atom to attract the bonding electrons in a covalent bond.

Pauling electronegativity values (see Figure 1) are used to compare the electronegativity of atoms of different elements.

- The greater the number, the greater the electronegativity.
- Fluorine is the most electronegative element.
- The closer an element is to fluorine, the more electronegative the element.

> **Synoptic link**
>
> For more details of covalent bonds, see Topic 5.3, Covalent bonding.

> **Key term**
>
> **Electronegativity:** The ability of an atom to attract the bonding electrons in a covalent bond.

electronegativity increases →

H 2.1							
Li 1.0	Be 1.5	B 2.0	C 2.5	N 3.0	O 3.5	F 4.0
Na 0.9	Mg 1.2	Al 1.5	Si 1.8	P 2.1	S 2.5	Cl 3.0
K 0.8						Br 2.8

▲ Figure 1 *Pauling electronegativity values*

Polar bonds

In a **polar bond**, the shared pair of electrons is shared unequally between the bonding atoms.

- The atom of the more electronegative element has a partial negative charge, shown by δ–.
- The atom of the less electronegative element has a partial positive charge, shown by δ+.

In a hydrogen chloride (HCl) molecule (Figure 2):

- the Cl atom is more electronegative and has a partial negative charge
- the H atom has a partial positive charge.

Dipoles

A **dipole** is the separation of partial charges in a molecule.

The separation of partial charges across a polar bond, arising from different electronegativities, is called a permanent dipole. Dipoles determine many properties of molecular compounds.

$\overset{\delta+}{H}-\overset{\delta-}{Cl}$

▲ Figure 2 *Hydrogen chloride, HCl*

> **Key terms**
>
> **Polar bond:** A covalent bond between atoms with different electronegativities with positive and negative partial charges on the bonded atoms.
>
> **Dipole:** Positive and negative partial charges separated by a short distance in a molecule.

> **Synoptic link**
>
> For details of different types of dipole in molecules, see Topic 6.3, Intermolecular forces.

31

Shapes of molecules and intermolecular forces

Polar and non-polar molecules

HCl has polar molecules – there is one polar bond, and there will be a dipole across the molecule.

For molecules with more than one polar bond, the dipoles in the polar bonds may cancel out due to their direction.

To decide whether a molecule is polar:

- Draw the 3D shape of the molecule.
- Label any polar bonds with partial charges, δ+ and δ–.
- If the molecule is symmetrical, the dipoles in the polar bonds cancel out, and the molecule is non-polar.
- If the molecule is not symmetrical, the dipoles in the polar bonds do not cancel out, and the molecule is polar.

Figure 3 shows how different molecular shapes result in H_2O having polar molecules and CO_2 having non-polar molecules.

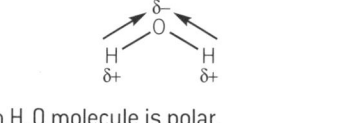

An H_2O molecule is polar.
There are two polar O—H bonds.
An H_2O molecule has a non-linear shape and the dipoles do **not** cancel.

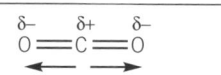

A CO_2 molecule is non-polar.
There are two polar C=O bonds, but a CO_2 molecule is symmetrical and the dipoles cancel.

▲ **Figure 3** *Non-polar and polar molecules*

Summary questions

1. What is meant by the term electronegativity? *(1 mark)*

2. Using Pauling values, show the partial charges on the following bonds.
   - a P–O
   - b Si–Cl
   - c N–S
   - d C=S
   (4 marks)

3. Explain, with reasons, whether the following molecules are polar or non-polar.
   - a NH_3
   - b BF_3
   - c CCl_4
   - d PH_3
   (8 marks)

6.3 Intermolecular forces

Specification reference: 2.2.2

Intermolecular forces are weak attractions that exist between molecules.

There are three types of intermolecular force:

- induced dipole–dipole interactions (London forces)
- permanent dipole–dipole interactions
- hydrogen bonds.

London forces (induced dipole–dipole interactions)

Electrons in a molecule are constantly moving:

- At any instant in time, the distribution of electrons may be uneven. As a result, a molecule may have a temporary dipole.

The dipoles can be represented using an arrow. The head of the arrow shows the region of negative charge.

The presence of a temporary dipole in one molecule can cause an induced dipole to form in a nearby molecule. The induced dipole can then induce a dipole in another neighbouring molecule.

The attraction between induced dipoles is called an induced dipole–dipole interaction or London force (Figure 1).

There are London forces between **all** molecules.

The effect on melting and boiling points

The more electrons a molecule has, the stronger the London forces between molecules. More energy will be needed to break the intermolecular forces, increasing the melting and boiling points. You can see this effect in Table 1, which compares the number of electrons with the boiling points of the noble gases.

Permanent dipole–dipole interactions

Polar molecules with a permanent dipole contain regions with different electron densities.

When two polar molecules are close, they will attract one another. This attraction, called a permanent dipole–dipole interaction, exists between any two molecules that have permanent dipoles, in addition to London forces.

For hydrogen chloride, the more electronegative Cl atom of one HCl molecule will attract the less electronegative H atom of another HCl molecule:

- In liquid HCl, these attractions are constantly breaking and re-forming as the molecules move around within the liquid.
- In solid HCl, these attractions hold the molecules in a fixed position.

> **Key term**
>
> **Intermolecular forces** are the weak attractive forces between molecules.

> **Synoptic link**
>
> For details of hydrogen bonds, see Topic 6.4, Hydrogen bonding.

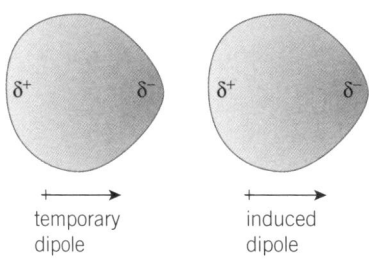

▲ **Figure 1** *The formation of induced dipole–dipole interactions*

▼ **Table 1** *Boiling points of the noble gases*

Noble gas	Number of electrons	Boiling point /K
He	2	1
Ne	10	25
Ar	18	84
Kr	36	116
Xe	54	161

> **Synoptic link**
>
> The role of London forces in melting and boiling points is covered further in Topic 8.2, The halogens, and Topic 12.1, Properties of the alkanes.

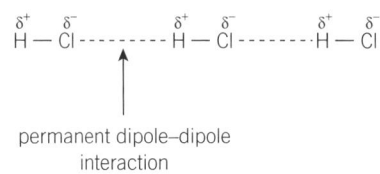

▲ **Figure 2** *Permanent dipole–dipole interactions*

Summary questions

1. How do permanent dipole–dipole interactions form between molecules? *(1 mark)*

2. Why does xenon have a higher boiling point than neon? *(2 marks)*

3. Explain how London forces form between molecules. *(2 marks)*

4. Why does F_2 have a higher boiling point than O_2, but a lower boiling point than HCl? *(2 marks)*

6.4 Hydrogen bonding
Specification reference: 2.2.2

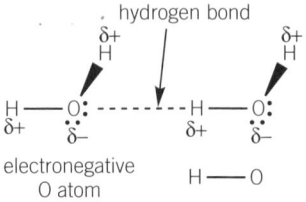

▲ Figure 1 *Hydrogen bonding between water molecules*

> **Revision tip**
> When drawing a diagram to show hydrogen bonding (Figure 1) you must include the following:
> - labelled dipoles on every molecule
> - lone pairs on the O, N, or F atom
> - a dotted line to represent the hydrogen bond.

> **Synoptic link**
> For details of hydrogen bonding in alcohols, see Topic 14.1, Properties of alcohols.

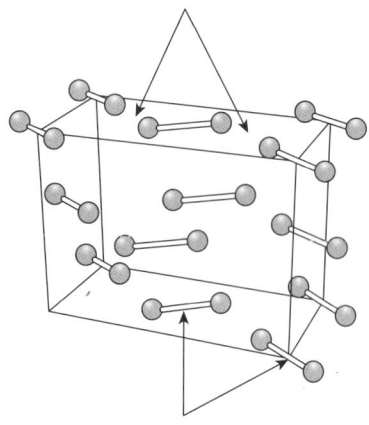

▲ Figure 2 *Simple molecular lattice of iodine, I_2*

Hydrogen bonds

Hydrogen bonds are especially strong permanent dipole–dipole interactions.

Hydrogen bonds occur between:

- a H atom bonded to an electronegative atom (O, N, or F) in one molecule
- a lone pair of an electronegative element (O, N, or F) in a different molecule.

Common examples of hydrogen bonding are in H_2O, NH_3 and HF.

Anomalous properties of H_2O
Ice is less dense than water

Ice has a very regular structure in which H_2O molecules are held in an open lattice structure by hydrogen bonds between the molecules, allowing ice to float on water. When ice melts, the open lattice structure collapses, allowing the H_2O molecules to move closer together and increasing the density.

H_2O has a much higher melting and boiling point than expected

Hydrogen bonds greatly increase the strength of the overall intermolecular forces in ice and water. Energy is needed to break the hydrogen bonds, increasing the melting and boiling point compared with similar molecules without hydrogen bonding.

Simple molecular lattices

In the solid state, a simple molecular lattice (e.g. I_2 in Figure 2) contains molecules held together in a regular structure:

- The atoms within each molecule are bonded together strongly by covalent bonds.
- The molecules are held in place by weak intermolecular forces.

Melting and boiling points

Simple molecular substances can be a solid, liquid, or gas at room temperature, but they tend to have relatively low melting points. The weak intermolecular forces are easily overcome at low temperatures.

Electrical conductivity

Simple molecules are poor electrical conductors, as they do not have a charge.

Solubility

In general, simple molecules have weak dipoles and tend to be soluble in non-polar solvents and insoluble in water. However, highly polar molecules are soluble in water because they can interact with the dipoles in water molecules.

> **Summary questions**
>
> 1 Why does ice float on water? *(2 marks)*
>
> 2 Why does water have a higher melting point than expected? *(2 marks)*
>
> 3 Draw labelled diagrams to show the hydrogen bonding between:
>   a HF molecules *(2 marks)*
>   b an H_2O molecule and an NH_3 molecule. *(2 marks)*

Chapter 6 Practice questions

1. Which molecule has the greatest bond angle?

   A BF_3 **B** SF_6 **C** NH_3 **D** H_2O *(1 mark)*

2. Which molecule is **not** polar?

   A CO_2 **B** H_2O **C** CH_3OH **D** NH_3 *(1 mark)*

3. Which compound does **not** have hydrogen bonding in the liquid state?

   A NH_3 **B** CH_4 **C** CH_3OH **D** HF *(1 mark)*

4. This question is about the shapes of four molecules, CO_2, H_2O, NH_3, and BH_3.

   a State the number of bonded pairs and lone pairs of electrons around the central atom. Ignore any inner shells.

Molecule	CO_2	H_2O	NH_3	BH_3
Bonded pairs				
Lone pairs				

   (4 marks)

   b For each molecule, name the shape and bond angles.

Molecule	CO_2	H_2O	NH_3	BH_3
Shape				
Bond angle				

   (4 marks)

5. The boiling points of argon, hydrogen chloride, hydrogen fluoride are shown below.

Compound	Ar	HCl	HF
Boiling point/°C	−186	−85	20
Number of electrons	10	18	18

   a All three molecules have London forces.

      Explain how London forces arise. *(3 marks)*

   b i Name the additional forces present in hydrogen chloride. *(1 mark)*

      ii Explain why HCl has these forces but Ar does **not**. *(3 marks)*

   c Name the additional forces present in hydrogen fluoride.

      Explain with a diagram how these forces arise. *(3 marks)*

   d Explain the trend in boiling points of these three compounds. *(2 marks)*

6. H has a Pauling electronegativity value of 2.1. Other values are shown below.

B	C	N	O	F
2.0	2.5	3.0	3.5	4.0
			P	Cl
			2.1	3.0
				Br
				2.8
				I
				2.5

   a What is meant by the term *electronegativity*? *(2 marks)*

   b Show, using δ+ and δ− symbols, the permanent dipoles on the following bonds.

      O—F;   O—I *(1 mark)*

   c Using the elements in the table, which bond will be the most polar and which will be non-polar? *(1 mark)*

   d Explain why CF_4 has polar bonds but non-polar molecules. *(3 marks)*

7.1 The periodic table

Specification reference: 3.1.1

Key terms

Period: A row of elements with their highest energy electron in the same shell.

Group: A column of elements with similar chemical and physical properties and with the same number of electrons in their outer shell.

Periodicity: The repeating trend in properties across each period in the periodic table.

Revision tip

Magnesium, Mg, has the electron configuration of $1s^2 2s^2 2p^6 \mathbf{3s^2}$.

Mg is in Group **2** because there are **2** electrons in the outer shell.

Mg is in Period **3** because the highest energy electron in the **3rd** shell.

Synoptic link

Electron configuration and blocks were first covered in Topic 5.1, Electron structure.

Summary questions

1. Define the term periodicity. *(1 mark)*

2. Give the full electron configuration for the elements below and classify the block of the element.
   a Boron b Fluorine
   c Neon   d Potassium
   e Iron.   *(10 marks)*

3. Predict the outer shell electron configuration for the following.
   a Rb ($Z = 37$)
   b Bi ($Z = 83$) *(2 marks)*

The structure of the periodic table

Arrangement of elements

Elements are arranged in order of increasing atomic number, with each successive element having one more proton.

Groups

Elements in the same group have the same number of electrons in their outer shell and have similar chemical and physical properties.

Periods

Elements in the same period (horizontal row) have their highest energy electrons in the same electron shell.

Periodicity

Periodicity is a repeating trend in properties across each period.

You will need to learn about periodic trends in electron configuration, ionisation energy, structure, and melting point.

Trend in electron configuration

Across a period, each successive element gains one more electron. Figure 1 shows the periodicity in outer shell electron configuration across Periods 2 and 3.

	Li	Be	B	C	N	O	F	Ne
Period 2	$2s^1$	$2s^2$	$2s^2 2p^1$	$2s^2 2p^2$	$2s^2 2p^3$	$2s^2 2p^4$	$2s^2 2p^5$	$2s^2 2p^6$
	Na	Mg	Al	Si	P	S	Cl	Ar
Period 3	$3s^1$	$3s^2$	$3s^2 3p^1$	$3s^2 3p^2$	$3s^2 3p^3$	$3s^2 3p^4$	$3s^2 3p^5$	$3s^2 3p^6$

▲ **Figure 1** *Periodicity in electron configuration*

- Looking down each group, there is the same number of electrons in each outer shell and type of sub-shell.
- Looking across each period, the only difference is the shell number.

Classification into blocks

The periodicity in electron configuration allows elements to be classified into blocks based on the sub-shell that contains the highest energy electrons. The s-, p-, and d-blocks are shown in Figure 2.

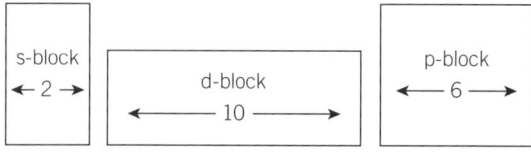

▲ **Figure 2** *Blocks of the periodic table*

- s-block elements have their highest energy electrons in an s sub-shell.
- p-block elements have their highest energy electrons in a p sub-shell.
- d-block elements have their highest energy electrons in a d sub-shell.

The width of each block is the same as the number of electrons that fill the sub-shell.

7.2 Ionisation energies
Specification reference: 3.1.1

Ionisation energy

Ionisation energies refer to the energy needed to remove electrons. The value of the ionisation energy depends on the attraction between the electron being lost and the nucleus. This is influenced by three factors:

Atomic radius: The greater the atomic radius, the smaller the attraction.

Nuclear charge: The greater the nuclear charge (number of protons), the greater the attraction.

Electron shielding: The greater the number of shells, the greater the shielding and the smaller the attraction.

First ionisation energy

First ionisation energy is the energy required to remove one electron from each atom in 1 mole of gaseous atoms to form 1 mole of gaseous 1+ ions. The equation for the first ionisation energy of magnesium is:

$$Mg(g) \rightarrow Mg^+(g) + e^-$$

Successive ionisation energies of magnesium

Electrons can be removed, one by one, from a gaseous atom to give successive ionisation energies. For example, the second ionisation energy of magnesium is the energy required to remove one electron from each ion in 1 mole of gaseous 1+ ions to form 1 mole of gaseous 2+ ions.

$$Mg^+(g) \rightarrow Mg^{2+}(g) + e^-$$

The value of a successive ionisation energy increases with ionisation number:

- As each electron is lost, there is the same number of protons attracting fewer electrons.
- The electrons are drawn in slightly closer to the nucleus, increasing the attraction.

Successive ionisation energies and shells

Successive ionisation energies provide evidence for the existence of shells as shown in Figure 1 for fluorine.

> **Synoptic link**
>
> Electron structure in terms of shells was introduced in Topic 5.1, Electron structure.

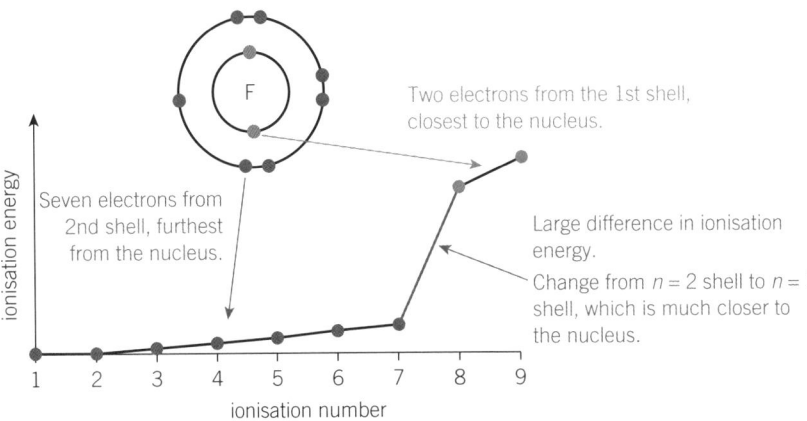

▲ **Figure 1** *Successive ionisation energies of fluorine*

Periodicity

Notice the large increase between the 7th and 8th ionisation energies of fluorine:

- The 8th electron is being removed from a different shell, closer to the nucleus.
- The 8th electron is attracted much more strongly by the nucleus than the 7th electron.

Successive ionisation energies and groups

Successive ionisation energies can be used to predict the group of an element:

- The successive ionisation energies of an element are compared.
- Any large increase shows that the next electron is being removed from a different shell, closer to the nucleus.

In Table 1, there is a large increase between the 3rd and 4th ionisation energies. The element would then have 3 electrons in its outer shell and be in Group 13 (3) of the periodic table.

You can see a similar increase for fluorine in Figure 1.

▼ **Table 1** *Successive ionisation energies of a Group 13 (3) element*

Ionisation number	Ionisation energy/kJ mol^{-1}
1st	578
2nd	1817
3rd	2745
4th	11 577
5th	14 842
6th	18 379

Trend in first ionisation energies down a group

First ionisation energy decreases down a group (See Table 2):

- Atomic radius increases as electrons are added to a different shell further from the nucleus.
- There are more inner shells, increasing electron shielding.
- There is less attraction between the nucleus and the outer electrons.

Down the group, the number of protons in the nucleus increases but the effects of atomic radius and shielding more than outweigh the increased nuclear charge.

▼ **Table 2** *First ionisation energies down Group 1*

Element	First ionisation energy /kJ mol^{-1}
Li	520
Na	496
K	419
Rb	403
Cs	376

Trend in first ionisation energies across a period

Figure 2 shows the trend in first ionisation energies for the first 20 elements in the periodic table.

General trend across a period

First ionisation energy shows a general increase across Periods 2 and 3:

- Nuclear charge increases.
- Electrons are added to same shell.
- There is more attraction between the nucleus and the outer electrons, decreasing the atomic radius.

Synoptic link

You will see how the trend in ionisation energy down a group affects reactivity in Topic 8.1, Group 2.

Revision tip

See the difference:

Down a group, first ionisation energy *decreases*.

Increased atomic radius and increased shielding are more important factors than the increased nuclear charge.

Across a period, first ionisation energy *increases*.

Increased nuclear charge is a more important factor than atomic radius and shielding.

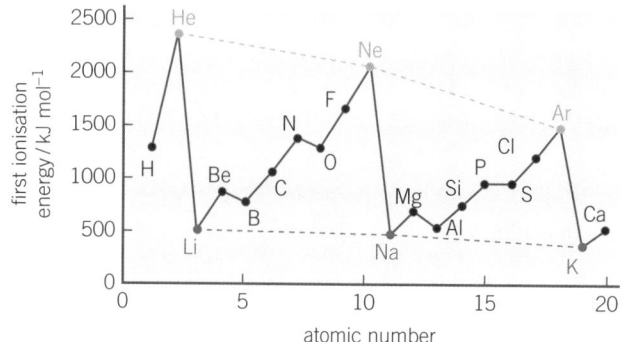

▲ **Figure 2** *First ionisation energies of first 20 elements*

First ionisation energy and sub-shells

Within the general increase in first ionisation energy across a period, there are two small falls:

- between Group 2 and 13 (3)   Period 2: Be → B   Period 3: Mg → Al
- between Group 15 (5) and 16 (6)   Period 2: N → O   Period 3: P → S

The fall between Group 2 and 13 (3)

Figure 3 compares the electron configurations of magnesium and aluminium.

Element	Electron configuration	'Electrons in boxes' diagrams				
		1s	2s	2p	3s	3p
Mg	$1s^2 2s^2 2p^6 3s^2$	↑↓	↑↓	↑↓ ↑↓ ↑↓	↑↓	
Al	$1s^2 2s^2 2p^6 3s^2 3p^1$	↑↓	↑↓	↑↓ ↑↓ ↑↓	↑↓	↑

▲ **Figure 3** *Electron configurations of magnesium and aluminium*

- In Al, the electron is lost from the 3p sub-shell which has a higher energy than the 3s sub-shell.
- The 3p electron in Al is lost more easily than a 3s electron in Mg.
- Therefore, Al has a smaller first ionisation energy than Mg.

The fall between Group 15 (5) and 16 (6)

Figure 4 compares the electron configurations of phosphorus and sulfur.

Element	Electron configuration	'Electrons in boxes' diagrams				
		1s	2s	2p	3s	3p
P	$1s^2 2s^2 2p^6 3s^2 3p^3$	↑↓	↑↓	↑↓ ↑↓ ↑↓	↑↓	↑ ↑ ↑
S	$1s^2 2s^2 2p^6 3s^2 3p^4$	↑↓	↑↓	↑↓ ↑↓ ↑↓	↑↓	↑↓ ↑ ↑

▲ **Figure 4** *Electron configurations of phosphorus and sulfur*

- In S, the electron is lost from the 3p orbital that contains paired electrons.
- In P, all three 3p orbitals contain one unpaired electron.
- In S, the paired electrons repel one another and one of the paired electrons is lost more easily than one of the unpaired electrons in P.
- Therefore, S has a lower first ionisation energy than P.

> **Synoptic link**
>
> Review Topic 5.1, Electron structure, for details about electron structure, shells, sub-shells, and orbitals.

> **Revision tip**
>
> The two small falls in first ionisation energy have been explained for Period 3.
>
> The explanation is essentially the same for the two small falls in Period 2.

Summary questions

1. Define the term 'first ionisation energy'. *(2 marks)*

2. Write the equation that represents the fourth ionisation energy of carbon. *(1 mark)*

3. State and explain the general trend in first ionisation energy across a period in the periodic table. *(3 marks)*

4. Why does oxygen have a lower first ionisation energy than nitrogen? *(2 marks)*

7.3 Periodic trends in bonding and structure

Specification reference: 3.1.1

> **Revision tip**
> Be careful not to confuse different types of lattice.
>
> An ionic lattice is a regular arrangement of positive and negative ions.
>
> A metallic lattice is a regular arrangement of positive ions surrounded by a sea of delocalised electrons.

Metals and non-metals

Across each period, there is a broad trend in the type of element, structure, and properties.

▼ **Table 1** Trends across a period of the periodic table

Type of element:	metals	→	non-metals		
Electrical conductivity	good	→	poor		
Structure:	giant metallic	→	giant covalent	→	simple molecular
Melting point	high			→	low

In each successive period, the changeover from metal to non-metal slowly shifts to the right.

Metallic bonding and structure

Metals have a giant metallic lattice structure held together by metallic bonds. In metallic bonding (see Figure 1):

- Each metal atom forms a positive ion (cation).
- The positive ions are arranged into a regular lattice structure.
- The outer shell electrons are delocalised into a 'sea of electrons', which can move throughout the structure.

As a result, the delocalised electrons are not attracted to any particular ion. They are free to move through the structure, allowing metals to conduct electricity, even in the solid state.

A metallic bond is the strong attraction between the positive ions in the lattice and the delocalised sea of electrons.

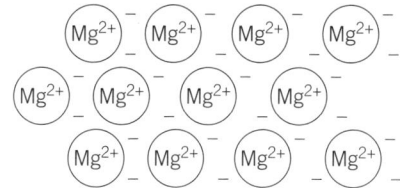

▲ **Figure 1** The metallic bonding in magnesium, with positive ions (Mg^{2+}) surrounded by delocalised electrons (−)

Giant covalent lattices

A giant covalent structure contains many billions of atoms held together by a network of strong covalent bonds. Across the periodic table, the first non-metals reached form giant structures. In Periods 2 and 3, boron, carbon, and silicon form giant covalent lattices.

A large amount of energy is needed to break the strong covalent bonds between the atoms. Therefore, substances with giant covalent structures have very high melting and boiling points.

Carbon and silicon

Carbon forms different lattices (diamond, graphite, and graphene), depending on how the atoms are arranged (see Figure 2). The different structures have different properties.

- Diamond and silicon form a giant 3D structure with atoms bonded in a tetrahedral arrangement.
- Graphite forms a giant planar structure with many planes weakly held together.
- Graphene is a single layer of graphite.

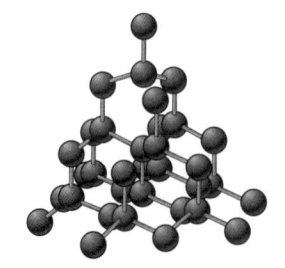

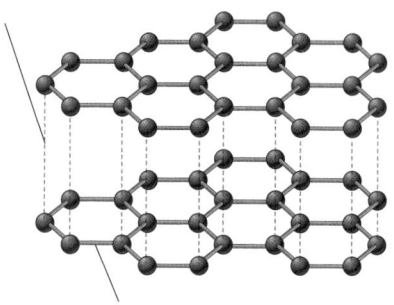

▲ **Figure 2** The structures of diamond (top) and graphite (bottom)

Electrical conductivity in carbon and silicon

Giant covalent lattices are non-conductors of electricity. The only exceptions are the graphene and graphite structures of carbon.

- In diamond and silicon, all four outer-shell electrons are involved in covalent bonding, so the electrons cannot move and are not available for conducting electricity.
- In graphite and graphene, three outer-shell electrons are involved in covalent bonding within each layer, with the other outer shell electron able to move and conduct electricity.

Trends in melting points across Periods 2 and 3

Trends in melting point are related to the change in bonding and structure, as shown in Figure 3.

Giant metallic lattices

Examples of giant metallic lattices are Li and Be (Period 2); Na, Mg, and Al (Period 3).

- The giant metallic lattice is held together by strong metallic bonds between positive ions and delocalised electrons.
- A comparatively large amount of energy is needed to break the metallic bonds and the melting points are high.
- The melting point increases from Li → Be and from Na → Mg → Al because the charge on the positive ion and number of delocalised electrons both increase. The attraction between the particles increases and more energy is needed to break the metallic bonds.

Giant covalent lattices

Examples of giant covalent lattices are B and C (Period 2); Si (Period 3).

- The lattice is held together by strong covalent bonds between atoms.
- A large amount of energy is needed to break the covalent bonds and the melting points are high.

Simple molecular lattices

Examples of simple molecular lattices are N_2, O_2, F_2, and Ne (Period 2); P_4, S_8, Cl_2, and Ar (Period 3).

- Weak London forces between molecules hold the lattice together.
- A small amount of energy is needed to break the London forces and the melting points are low.
- In Figure 3, notice the fluctuations in the melting points of P_4, S_8, and Cl_2. The London forces of attraction increase with the number of electrons in the molecules. As a result, the melting point increases in the order: $Ar < Cl_2 < P_4 < S_8$.

> **Synoptic link**
>
> For details of simple molecular lattices, see Topic 6.4, Hydrogen bonding.

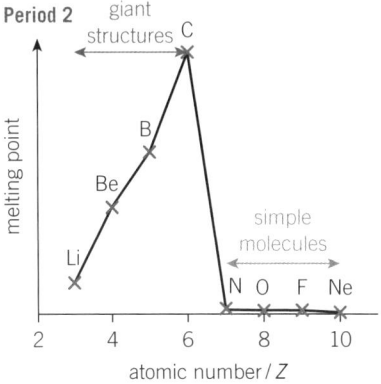

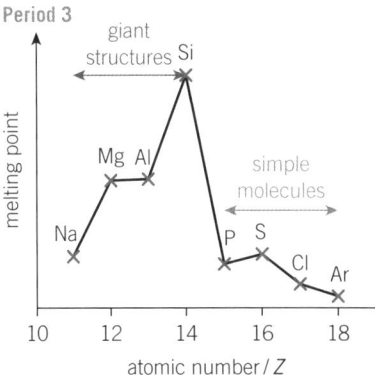

▲ **Figure 3** *Melting points of the Period 3 elements*

> **Synoptic link**
>
> For details of the simple molecular lattice structure, see Topic 6.4, Hydrogen bonding.

Summary questions

1. What is the bonding and structure in the lattices for:
   a chlorine   b sodium? *(4 marks)*

2. State and explain the trend in electrical conductivity across Period 3. *(3 marks)*

3. Explain why sulfur melts at 388 K but silicon melts at 1687 K. *(4 marks)*
4. Explain why phosphorus has a higher melting point than chlorine. *(2 marks)*

Chapter 7 Practice questions

1. Which element has the highest first ionisation energy?

   A Li   **B** F   **C** Ne   **D** K   *(1 mark)*

2. Which equation represents the second ionisation energy of magnesium?

   A $Mg(g) \rightarrow Mg^{2+}(g) + 2e^-$ **B** $Mg^+(g) \rightarrow Mg^{2+}(g) + e^-$

   C $Mg^{2+}(g) + e^- \rightarrow Mg^{3+}(g)$ **D** $Mg^-(g) + e^- \rightarrow Mg^{2-}(g)$ *(1 mark)*

3. The first eight ionisation energies, in kJ mol⁻¹, of an element in Period 3 are listed below

   789, 1577, 3232, 4356, 16091, 19785, 23787, 29253.

   What is the element?

   A Na   **B** Mg   **C** Al   **D** Si   *(1 mark)*

4. What is the order of increasing melting point for the elements lithium, boron, carbon, and nitrogen?

   smallest ⇒ largest

   A Li < B < C < N   **B** C < Li < N < B

   C N < Li < B < C   **D** B < N < C < Li   *(1 mark)*

5. Which structure is planar?

   A sodium   **B** silicon

   C iodine   **D** graphene   *(1 mark)*

6. Table 1 shows melting points across Period 3.

   a Which element(s) have

       i a giant metallic structure? *(1 mark)*

       ii a giant covalent structure? *(1 mark)*

       iii a simple molecular structure? *(1 mark)*

   b This trend in melting points is similar across Period 2. What name is given to a repeating trend across periods? *(1 mark)*

   c State what is meant by metallic bonding.

       Use a diagram as part of your answer. *(2 marks)*

   d Explain why the boiling point of phosphorus is much lower than that of silicon. *(3 marks)*

   e Explain why sulfur has a higher melting point than chlorine. *(2 marks)*

   f Predict the trend in electrical conductivity of the solid elements across Period 3. Explain the trend for the metals. *(3 marks)*

▼ Table 1

element	melting point / °C
Na	98
Mg	639
Al	660
Si	1410
P	44
S	113
Cl	−101

7. Figure 1 shows the variation in first ionisation energies of elements across Period 2 of the periodic table.

   a Write an equation, with state symbols, to represent the first ionisation energy of carbon. *(1 mark)*

   b Explain why there is a general increase in first ionisation across Period 2. *(3 marks)*

   c Explain why the first ionisation energy of boron is less than that of beryllium. *(2 marks)*

   d Explain why the first ionisation energy of oxygen is less than that of nitrogen. *(2 marks)*

   e Sodium has 11 successive ionisation energies.

       i Write an equation, including state symbols, for the 7th ionisation energy of sodium. *(1 mark)*

       ii Why do the successive ionisation energies of sodium increase in value? *(1 mark)*

       iii Explain how the successive ionisation energies of sodium provide evidence for its electron structure. *(3 marks)*

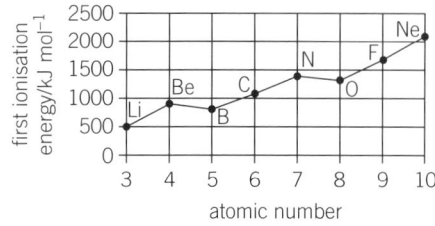

▲ Figure 1

8.1 Group 2
Specification reference: 3.1.2

The Group 2 metals
The Group 2 metals are elements in the s-block of the periodic table. The atoms have two electrons in their outer shell, s².

The Group 2 metals are all reactive metals, and have relatively low densities. In redox reactions, each atom is oxidised, losing two electrons to form a positive ion with a 2+ charge.

e.g. $Mg \rightarrow Mg^{2+} + 2e^-$

The outer shell electron configurations for Group 2 atoms and their 2+ ions are shown in Table 1.

▼ **Table 1** *Electron configuration of Group 2 atoms and their 2+ ions*

Group 2 atom		Group 2 ion	
Be	[He] 2s²	Be²⁺	[He]
Mg	[Ne] 3s²	Mg²⁺	[Ne]
Ca	[Ar] 4s²	Ca²⁺	[Ar]
Sr	[Kr] 5s²	Sr²⁺	[Kr]
Ba	[Xe] 6s²	Ba²⁺	[Xe]
Ra	[Rn] 7s²	Ra²⁺	[Rn]

Ionisation energies and reactivity
Reactivity increases down Group 2, largely because of ionisation energies.

Ionisation energies
The Group 2 metals react by losing two electrons, which requires the first two ionisation energies.

The equations for the first and second ionisation energies of magnesium are shown below:

First ionisation energy $Mg(g) \rightarrow Mg^+(g) + e^-$

Second ionisation energy $Mg^+(g) \rightarrow Mg^{2+}(g) + e^-$

The first and second ionisation energies both decrease down a group. (See Table 2.)

- Atomic radius increases as electrons are added to a different shell further from the nucleus.
- More inner shells, increasing electron shielding.
- Less attraction between the nucleus and the outer electrons.

Down the group, the number of protons in the nucleus increases but the effects of atomic radius and shielding more than outweigh the increased nuclear charge.

Trend in reactivity
The reactivity increases down Group 2.

- The total ionisation energy to remove two electrons decreases down the group.
- Electrons are lost more easily and gained by other species.
- The elements become stronger reducing agents.

Redox reactions
In their redox reactions, the Group 2 metals act as strong reducing agents, adding electrons easily to other substances.

Redox reaction with oxygen
The Group 2 metals all react with oxygen to form a metal oxide, e.g.

$$2Mg(s) + O_2(g) \rightarrow 2MgO(s)$$

0 → +2   oxidation
0 → −2   reduction

The reactivity of Group 2 metals with oxygen increases down the group.

> **Synoptic link**
>
> Ionisation energies are covered in detail in Topic 7.2, Ionisation energies.

▼ **Table 2** *The first and second ionisation energies of Group 2 metals*

	Ionisation energy/ kJmol⁻¹		
Element	1st	2nd	1st + 2nd
Mg	738	1450	2188
Ca	590	1145	1735
Sr	550	1064	1614
Ba	503	965	1468

> **Synoptic link**
>
> For more details of redox reactions, see Topic 4.3, Redox.

> **Key term**
>
> **Reducing agent:** A substance that adds electrons to other species and loses electrons itself.

Reactivity trends

Redox reaction with water

Group 2 metals all react with water to form a metal hydroxide and hydrogen gas. Magnesium reacts very slowly, but the reaction becomes much more vigorous with metals further down the group.

$$Ca(s) + 2H_2O(l) \rightarrow Ca(OH)_2(aq) + H_2(g)$$

| 0 | → +2 | oxidation |
| +1 | → 0 | reduction |

The reactivity of Group 2 metals with water increases down the group.

Redox reaction with dilute acids

The Group 2 metals all react with dilute acids to form a salt and hydrogen gas.

$$\text{metal} + \text{acid} \rightarrow \text{salt} + \text{hydrogen}$$
$$Mg(s) + 2HCl(aq) \rightarrow MgCl_2(aq) + H_2(g)$$

| 0 | → +2 | oxidation |
| +1 | → 0 | reduction |

The reactivity of Group 2 metals with dilute acids increases down the group.

Reactions of Group 2 compounds

Group 2 oxides

The Group 2 oxides react with water forming alkaline solutions of Group 2 hydroxides.

$$CaO(s) + H_2O(l) \rightarrow Ca(OH)_2(aq)$$

The Group 2 hydroxides are only slightly soluble in water and much of the hydroxide product may form a precipitate instead of an alkaline solution.

The solubility of the Group 2 hydroxides in water increases down the group. The solutions become more alkaline because they contain more $OH^-(aq)$ ions.

Uses of Group 2 compounds

The Group 2 oxides, hydroxides, and carbonates are bases. They are widely used to neutralise acids.

- Calcium hydroxide, $Ca(OH)_2$, is used in agriculture to neutralise acid soils, e.g.
  $$Ca(OH)_2(s) + 2H^+(aq) \rightarrow Ca^{2+}(aq) + 2H_2O(l)$$

- Magnesium hydroxide, $Mg(OH)_2$, and calcium carbonate, $CaCO_3$, are used as 'antacids' to neutralise acids that cause indigestion (e.g. HCl).
  $$Mg(OH)_2(s) + 2HCl(aq) \rightarrow MgCl_2(aq) + 2H_2O(l)$$
  $$CaCO_3(s) + 2HCl(aq) \rightarrow CaCl_2(aq) + H_2O(l) + CO_2(g)$$

> **Synoptic link**
>
> For more details of reactions of metals with acids, see Topic 4.3, Redox.

Summary questions

1. State and explain the trend in first ionisation energy of the Group 2 metals. *(3 marks)*

2. Write an equation for the second ionisation energy of strontium. *(1 mark)*

3. Write an equation, with state symbols, for the reaction between barium and water.
   Use oxidation numbers to explain which element is oxidised and which is reduced. *(3 marks)*

8.2 The halogens
Specification reference: 3.1.3

The halogens
The halogens are the elements in Group 17 (Group 7) of the p-block of the periodic table. Halogen atoms have seven electrons in their outer shell, s^2p^5.

The halogens are all reactive non-metals. In many redox reactions, each halogen atom is reduced, gaining one electron to form a halide ion with a 1− charge.

e.g. $Cl_2 + 2e^- \rightarrow 2Cl^-$

The outer shell electron configurations for halogen atoms and halide ions are shown in Table 1.

The physical properties of halogens
The boiling points of the halogens increase down the group.

- The halogens exist as diatomic molecules, e.g. Cl_2, Br_2, I_2.
- In the solid state, the halogens form simple molecular lattices, with weak London forces between the halogen molecules.
- Down the group, the halogen molecules contain more electrons, increasing the strength of the London forces.
- More energy is required to break the London forces, increasing the boiling point.

Redox reactions
In their redox reactions, the halogens act as strong oxidising agents, removing electrons easily from other substances. In their redox reactions, halogens are often reduced to form halide 1− ions.

Trend in reactivity
To form a 1− ion, a halogen atom must attract an electron from another species and place the electron in its outer shell.

Down the group:

- The number of shells increases and the atomic radius and shielding increases.
- The attraction between the nucleus and the outer shell decreases.
- The attraction for an external electron decreases and the oxidising ability decreases.

Redox reactions between halogens and halides
The redox reactions between halogens and aqueous halides can be used to show that the reactivity of halogens decreases down the group.

- A solution of each halogen is mixed with solutions of the other halides.
- If a reaction takes place, the halogen oxidises the halide ion, removing an electron from each halide ion.

Solutions of different halogens have different colours. These colours can be used to show whether a reaction has taken place. Aqueous solutions of bromine and iodine have similar colours. An organic solvent, such as cyclohexane, is added and the mixture shaken. The halogen dissolves in the organic solvent to produce colours that are much easier to tell apart, particularly for bromine and iodine (See Table 2).

▼ **Table 1** *Electron configuration of halogen atoms and halide ions*

Halogen atom		Halide ion	
F	$2s^2 2p^5$	F⁻	$2s^2 2p^6$
Cl	$3s^2 3p^5$	Cl⁻	$3s^2 3p^6$
Br	$4s^2 4p^5$	Br⁻	$4s^2 4p^6$
I	$5s^2 5p^5$	I⁻	$5s^2 5p^6$
At	$6s^2 6p^5$	At⁻	$6s^2 6p^6$

Revision tip
At room temperature:
- Fluorine is a yellow gas.
- Chlorine is a pale green gas.
- Bromine is a red-brown liquid.
- Iodine is a dark grey solid.

Synoptic link
For more details about London forces look back at Topic 6.3, Intermolecular forces.

Synoptic link
For more details of redox reactions, see Topic 4.3, Redox.

Key term
Oxidising agent: A substance that removes electrons from other species and gains electrons itself.

Revision tip
Reactivity decreases down the halogens group:
- Electrons are gained less easily from other species.
- The elements become weaker oxidising agents.

▼ **Table 2** *Colours of halogens in water and cyclohexane*

Halogen	Water	Cyclohexane
Chlorine, Cl_2	pale green	pale green
Bromine, Br_2	orange	orange
Iodine, I_2	brown	violet

Reactivity trends

The results of the experiment confirm that reactivity decreases down the group:

- Cl_2 reacts with Br^- and I^-.
- Br_2 reacts with I^- only.
- I_2 does not react.

In the reaction of Cl_2 and Br^-, chlorine oxidises bromide ions.

$$Cl_2(aq) + 2Br^-(aq) \rightarrow 2Cl^-(aq) + Br_2(aq)$$

0	$\rightarrow$ −1	reduction
−1	$\rightarrow$ 0	oxidation

Half-equations: $Cl_2(aq) + 2e^- \rightarrow 2Cl^-(aq)$ reduction

$2Br^-(aq) \rightarrow Br_2(aq) + 2e^-$ oxidation

Similar equations and oxidation number changes can be written for the reactions of Cl_2 with I^-; Br_2 with I^-.

Disproportionation reactions of chlorine

Disproportionation is the simultaneous oxidation and reduction of the same element.

Chlorine and water treatments

Chlorine reacts with water in a redox reaction to form two acids: chloric(I) acid (HClO) and hydrochloric acid (HCl).

$$Cl_2(aq) + H_2O(l) \rightarrow HClO(aq) + HCl(aq)$$

0	$\rightarrow$ +1	oxidation
0	$\rightarrow$ −1	reduction

This is an example of disproportionation. Chlorine is simultaneously oxidised (from 0 to +1 in HClO) and reduced (from 0 to −1 in HCl).

This reaction is used in water treatment to kill bacteria that could be harmful for human health. Care is needed because chlorine is toxic and can react with hydrocarbons (e.g. CH_4) to form chlorinated hydrocarbons, which are carcinogenic (cause cancer).

The reaction of chlorine with sodium hydroxide

Chlorine reacts with cold, dilute sodium hydroxide to form NaCl, sodium chlorate(I) (NaClO), and H_2O.

Sodium chlorate(I) is used in household bleach.

$$Cl_2(aq) + 2NaOH(aq) \rightarrow NaClO(aq) + NaCl(aq) + H_2O(l)$$

0	$\rightarrow$ +1	oxidation
0	$\rightarrow$ −1	reduction

This is another example of a disproportionation. Chlorine is simultaneously oxidised (from 0 to +1 in NaClO) and reduced (from 0 to −1 in NaCl).

Key term

Disproportionation: The simultaneous oxidation and reduction of the same element.

Summary questions

1. Write the electron configuration for a Br atom and a Br^- ion. *(2 marks)*

2. State and explain the trend in reactivity of the halogens down the group. *(3 marks)*

3. A chemist bubbles chlorine, Cl_2, gas through a solution of potassium iodide, KI.
   a Write the equation for the reaction. *(2 marks)*
   b State the colour that would be observed when cyclohexane is added and the reaction mixture is shaken. *(1 mark)*
   c Use oxidation numbers to explain which element is oxidised and which is reduced. *(2 marks)*

8.3 Qualitative analysis
Specification reference: 3.1.4

Identifying carbonate ions, CO_3^{2-}

To test for the presence of carbonate ions, CO_3^{2-}, in a solid or in solution:

- Add dilute nitric acid to the sample.
- If bubbles are observed, bubble the gas through limewater.
- If the limewater turns cloudy (milky), the gas is carbon dioxide and CO_3^{2-} ions are present.

The equation for the reaction with solid calcium carbonate is shown below:

$$CaCO_3(s) + 2HNO_3(aq) \rightarrow Ca(NO_3)_2(aq) + CO_2(aq) + H_2O(l)$$

Synoptic link
All carbonates react with acids in a similar way, as described in Topic 4.1, Acids, bases, and neutralisation.

Identifying sulfate ions, SO_4^{2-}

To test for the presence of sulfate ions, SO_4^{2-}, in an aqueous solution:

- Add aqueous barium chloride, $BaCl_2(aq)$, or barium nitrate, $Ba(NO_3)_2(aq)$.
- If a white precipitate of barium sulfate is formed, SO_4^{2-} ions are present:

$$Ba^{2+}(aq) + SO_4^{2-}(aq) \rightarrow BaSO_4(s)$$

Identifying halide ions, Cl^-, Br^-, or I^-

To test for the presence of halide ions in an aqueous solution:

- Add aqueous silver nitrate, $AgNO_3(aq)$, to the solution.
- If a precipitate is formed, the solution contains halide ions.

The equation for the reaction of aqueous chloride ions is shown below.

$$Ag^+(aq) + Cl^-(aq) \rightarrow AgCl(s)$$

The colour of the precipitate identifies the halide, as shown in Table 1.

It is sometimes difficult to identify the colour of the precipitate, and its solubility in aqueous ammonia is used to identify the precipitate (Table 2).

▼ **Table 1** *Identifying halide ions*

Halide ion	Precipitate formed	Colour of precipitate
Chloride	AgCl	white
Bromide	AgBr	cream
Iodide	AgI	yellow

▼ **Table 2** *Solubility of silver halides in ammonia*

Silver halide	Solubility in ammonia	
	Dilute	Conc.
AgCl	Yes	Yes
AgBr	No	Yes
AgI	No	No

Mixtures of ions

To analyse a mixture of ions you must carry out the tests in the order:

Carbonate → Sulfate → Halide

If you do not do this, you may observe conflicting results:

- Carbonate ions produce a precipitate in the sulfate and halide tests.
- Sulfate ions produce a precipitate with the halide test.

Identifying ammonium ions

To test for the presence of ammonium ions, NH_4^+, in solution:

- Add NaOH(aq) to the solution and warm the mixture.
- Test for any gas evolved with damp red litmus paper.
- If the litmus paper turns blue, the gas is ammonia and NH_4^+ ions are present.

$$NH_4^+ + OH^- \rightarrow NH_3 + H_2O$$

Summary questions

1. Write the ionic equation for the reaction between barium ions and sulfate ions. Include state symbols. *(1 mark)*

2. How could you identify bromide ions? *(2 marks)*

3. How could you identify ammonium ions? *(2 marks)*

Chapter 8 Practice questions

1 Which property decreases down Group 2?
  A reactivity
  B second ionisation energy
  C pH when oxides are added to water
  D number of electrons in outer shell *(1 mark)*

2 Which equation shows the reaction of strontium with water?
  A $Sr + H_2O \rightarrow SrO + H_2$
  B $2Sr + H_2O \rightarrow Sr_2O + H_2$
  C $2Sr + 2H_2O \rightarrow 2SrOH + H_2$
  D $Sr + 2H_2O \rightarrow Sr(OH)_2 + H_2$ *(1 mark)*

3 Which equation does **not** show a disproportionation reaction?
  A $Cl_2 + 2OH^- \rightarrow Cl^- + ClO^- + H_2O$
  B $ClO_3^- + Cl^- + 6H^+ \rightarrow 3Cl_2 + 3H_2O$
  C $Cl_2 + H_2O \rightarrow HClO + HCl$
  D $4KClO_3 \rightarrow 3KClO_4 + KCl$ *(1 mark)*

4 Which silver halide dissolves in concentrated ammonia solution but **not** in dilute ammonia solution?
  A AgF   B AgCl   C AgBr   D AgI *(1 mark)*

5 Calcium is reacted with oxygen to form compound **A**. When **A** is added to water a solution forms, containing compound **B**. Compound **B** is used in agriculture.
  a Name compounds **A** and **B**.
    Write equations, with state symbols, for the reactions described.
    (4 marks)
  b Suggest the pH of the solution of compound **B**. *(1 mark)*
  c Why is compound **B** used in agriculture? Write an equation to illustrate your answer. *(2 marks)*

6 The reactivity of the Group 2 elements increases down the group.
  a Write equations for the reactions of barium with oxygen and water.
    (2 marks)
  b The increased reactivity down the group can be explained in terms of different ionisation energies.
    i Write an equation, with state symbols, to represent the 2nd ionisation energy of strontium. *(1 mark)*
    ii Explain why the ionisation energies change down Group 2 and how this helps to explain the trend in reactivity. *(4 marks)*
  c A student reacts 1.00 g of calcium and 1.00 g of magnesium with an excess of dilute hydrochloric acid. The student thought that they would obtain the same volume of gas from each reaction.
    i Write the equation, with state symbols, for the reaction of magnesium with dilute hydrochloric acid. *(2 marks)*
    ii Explain, with a calculation, whether the student was correct.
    (4 marks)

9.1 Enthalpy changes
Specification reference: 3.2.1

Exothermic and endothermic reactions

Enthalpy, H, is a measure of the heat energy stored in a chemical system.

In a reaction, there is often a difference between the enthalpy of the reactants and enthalpy of the products. This difference is called the enthalpy change, ΔH.

Exothermic reactions

In an exothermic reaction, the products have less enthalpy than the reactants:

- Chemical energy is changed into thermal (heat) energy.
- The chemicals lose energy.
- The energy that is lost by the chemicals is gained by the surroundings, which increase in temperature.

As a result, the enthalpy change (ΔH) for an exothermic reaction is negative.

> **Key terms**
>
> **Enthalpy (H):** The heat energy stored in a chemical system.
>
> **Enthalpy change (ΔH):** The heat energy change measured under conditions of constant pressure.

> **Key term**
>
> **Exothermic reaction:** Chemicals lose energy to the surroundings.
>
> ΔH for exothermic reactions is negative.

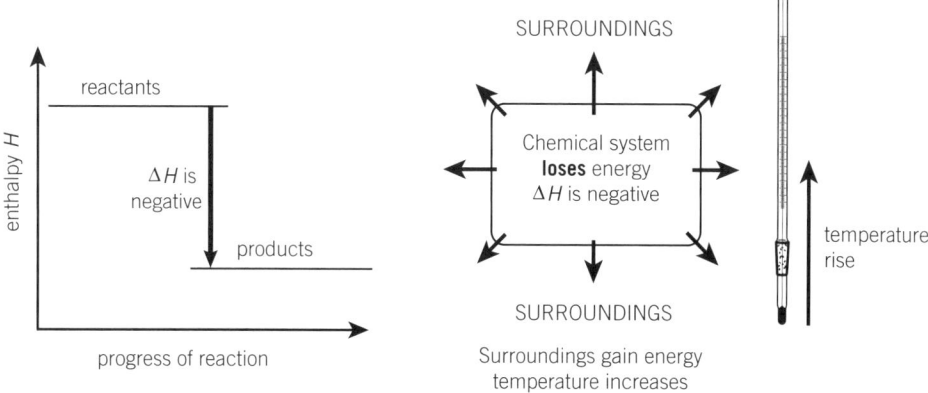

▲ **Figure 1** *Exothermic enthalpy profile diagram*

Endothermic reactions

In an endothermic reaction, the products have more enthalpy than the reactants:

- Thermal (heat) energy is changed into chemical energy.
- The chemicals gain energy.
- The energy that is gained by the chemicals is lost by the surroundings, which decrease in temperature.

As a result, the enthalpy change (ΔH) for an endothermic reaction is positive.

> **Key term**
>
> **Endothermic reaction:** Chemicals gain energy from the surroundings.
>
> ΔH for endothermic reactions is positive.

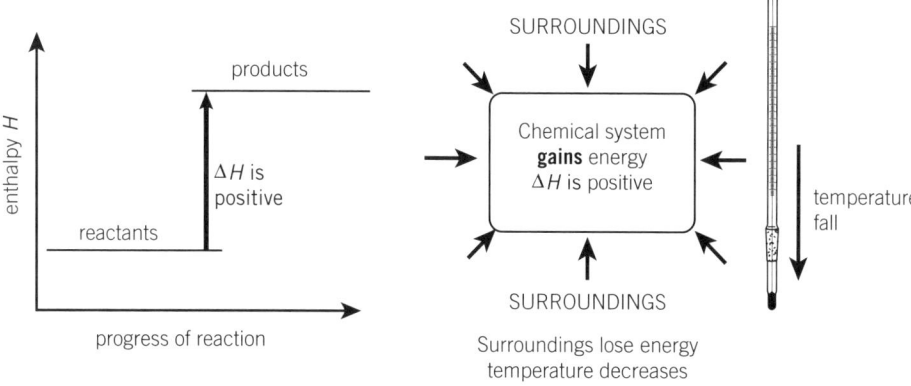

▲ **Figure 2** *Endothermic enthalpy profile diagram*

Enthalpy

Standard enthalpy changes

Enthalpy changes depend on temperature and pressure.

Standard enthalpy changes are measured using the same standard conditions.

- Any gases must have a pressure of 100 kPa.
- A stated temperature must be used, normally 298 K (25 °C).
- Any solutions must have a concentration of 1.00 mol dm^{-3}.

All substances must be in their standard states (their physical states under standard conditions).

Standard enthalpy change of reaction, $\Delta_r H$

The standard enthalpy change of reaction is the enthalpy change that accompanies a reaction in the molar quantities shown in a chemical equation under standard conditions, with all reactants and products in their standard states.

$\Delta_r H$ for the reaction of sodium and chlorine shown in the equation below, is:

$$2Na(s) + Cl_2(g) \rightarrow 2NaCl(s) \quad \Delta_r H = -822.4 \text{ kJ mol}^{-1}$$

2 mol   1 mol → 2 mol

Standard enthalpy change of formation, $\Delta_f H$

The standard enthalpy change of formation is the enthalpy change when one mole of a compound is formed from its elements in their standard states, under standard conditions.

$\Delta_f H$ for aluminium oxide, Al_2O_3, is represented as:

$$2Al(s) + 1\tfrac{1}{2}O_2(g) \rightarrow Al_2O_3(s) \quad \Delta_f H = -1675.7 \text{ kJ mol}^{-1}$$

elements → 1 mol

The standard enthalpy change of formation of an element in its standard state is zero because there is no chemical or physical change.

Standard enthalpy change of combustion, $\Delta_c H$

The standard enthalpy change of combustion is the enthalpy change when one mole of a compound is completely burnt in oxygen under standard conditions.

$\Delta_c H$ for methane, CH_4, is represented as:

$$CH_4(g) + 2O_2(g) \rightarrow CO_2(g) + 2H_2O(l) \quad \Delta_c H = -890.3 \text{ kJ mol}^{-1}$$

1 mol   → combustion products

Standard enthalpy change of neutralisation, $\Delta_{neut} H$

The standard enthalpy change of neutralisation is the energy change that accompanies the reaction of an acid by a base to form one mole of water, under standard conditions, with all reactants and products in their standard states.

$$H^+(aq) + OH^-(aq) \rightarrow H_2O(l) \quad \Delta_{neut} H = -57 \text{ kJ mol}^{-1}$$

→ 1 mol

Summary questions

1. What are standard conditions? *(1 mark)*

2. What is the value of $\Delta_f H$ for $N_2(g)$? *(1 mark)*

3. What is the equation that represents the standard enthalpy change of combustion of ethanol? Include state symbols. *(2 marks)*

4. What is the equation that represents the standard enthalpy change of formation of methane? Include state symbols. *(2 marks)*

9.2 Measuring enthalpy changes
Specification reference: 2.1.3, 3.2.1

Calculating an energy change

In experiments, we measure the heat energy change in the surroundings.

The heat energy change q, in joules, is given by the equation:

$$q = mc\Delta T$$

m = mass of the surroundings (g)

c = specific heat capacity of the surroundings ($J\,g^{-1}\,K^{-1}$)

ΔT = temperature change (final temperature – initial temperature)

> **Worked example: Using $q = mc\Delta T$**
>
> 50.0 g of water ($c = 4.18\,J\,g^{-1}\,K^{-1}$) are heated from 21.0 °C to 31.0 °C.
>
> $q = mc\Delta T = 50.0 \times 4.18 \times (31.0 - 21.0) = 2090\,J$

Revision tip

Using these units, $q = mc\Delta T$ gives a value for the energy change in joules (J).

To give an answer in kilojoules (kJ), divide by 1000.

Determination of an enthalpy change of a reaction

The enthalpy change of a reaction, in $kJ\,mol^{-1}$, is determined as the enthalpy change linked to a stated equation where the species are in the molar quantities shown by the balancing numbers in the equation.

> **Worked example: Calculating an enthalpy change of reaction from experimental results**
>
> A student adds 25.0 cm³ of 2.00 mol dm⁻³ sodium hydroxide solution into a polystyrene cup and adds 25.0 cm³ of 2.00 mol dm⁻³ hydrochloric acid. The initial temperature of both solutions is 22.5 °C. The final temperature of the reaction mixture is 36.5 °C.
>
> The mixture has a specific heat capacity, c, of $4.18\,J\,g^{-1}\,K^{-1}$ and a density of $1.00\,g\,cm^{-3}$.
>
> Calculate the enthalpy change of reaction:
>
> $NaOH(aq) + HCl(aq) \rightarrow NaCl(aq) + H_2O(l)$
>
> **Step 1:** Calculate the energy, q transferred with the surroundings (in kJ):
>
> $q = mc\Delta T = 50.0 \times 4.18 \times 14.0 = 2926\,J = 2.926\,kJ$
>
> **Step 2:** Calculate the amount, n, in mol:
>
> $n(NaOH) = n(HCl) = 2.00 \times \dfrac{25.0}{1000} = 0.0500\,mol$
>
> **Step 3:** Calculate the energy transfer for the moles in the equation (q/n):
>
> Energy transfer $= \dfrac{2.926}{0.0500} = 58.52\,kJ\,mol^{-1}$
>
> **Step 4:** Decide on sign for ΔH and show to 3 s.f. (the least accurate measurement):
>
> $\Delta_r H = -58.5\,kJ\,mol^{-1}$

Revision tip

To calculate the enthalpy change of a reaction:

- Calculate q using $q = mc\Delta T$
  Here $m = 50.0\,(25.0 + 25.0)$
- Convert to kJ
- Calculate the amount, n, in moles.
- Divide q by n.
- Add sign for ΔH.

Revision tip

Remember to add the sign for the enthalpy change.

Here, the surroundings increase in temperature, the reaction is exothermic and so ΔH has a negative sign.

Errors

The major source of error in this experiment is:

- heat loss to the surroundings.

Methods for reducing heat loss include:

- adding a lid to the polystyrene cup
- adding insulation around the polystyrene cup.

Revision tip

In the calculation, you can convert to kJ at **Step 1** or at a later step.

Just remember to carry out the conversion.

Enthalpy

Determination of an enthalpy change of combustion, ΔH_c

The enthalpy change of combustion, in kJ mol^{-1}, can be determined by measuring the energy change produced by burning a known mass of a fuel.

Synoptic link

For Step 2, you need to use the relationship between moles and mass, $n = m/M$. See Topic 3.1, Amount of substance and the mole.

Revision tip

You will always get the most accurate result by rounding numbers just once, at the final stage of a calculation. This is easy to do with modern calculators.

If you round each stage, you will be introducing rounding errors.

In this example, if you were to round q to 3 s.f., the final answer is -876 kJ mol^{-1} and you have introduced a rounding error.

Worked example: Calculating an enthalpy change of combustion from experimental results

A student added 100.0 cm^3 of water to a beaker. The initial temperature of the water was 18.5 °C. The student then lit a spirit burner and burnt 1.15 g of ethanol. The final temperature of the water was 71.0 °C.

Water has a specific heat capacity, c, of 4.18 J g^{-1} K^{-1} and a density of 1.00 g cm^{-3}.

Calculate the enthalpy change of combustion of ethanol:

Step 1: Calculate the energy, q, transferred with the surroundings (in kJ):

$$q = mc\Delta T = 100.0 \times 4.18 \times 52.5 = 21\,945\text{ J} = 21.945\text{ kJ}$$

Step 2: Calculate the amount, n, in moles of ethanol that is burnt.

$$n(C_2H_5OH) = \frac{1.15}{46.0} = 0.0250 \text{ mol}$$

Step 3: Calculate the energy transfer for 1 mol of ethanol (q/n).

$$\text{Energy transfer} = \frac{21.945}{0.0250} = 877.8 \text{ kJ mol}^{-1}$$

Step 4: Decide on sign for ΔH and show to 3 s.f. (the least accurate measurement):

$$\Delta_c H = -878 \text{ kJ mol}^{-1}$$

Errors

The major sources of error in this experiment are:
- heat loss to the surroundings
- incomplete combustion of ethanol
- evaporation of ethanol from the wick
- non-standard conditions being used.

Methods for reducing heat loss include:
- adding a lid to the beaker
- using draft shields around the apparatus

Summary questions

1. What is the major source of error in an enthalpy change experiment? How could this error be reduced? *(2 marks)*

2. A student adds 50.0 cm^3 of 1.00 mol dm^{-3} copper sulfate solution to a polystyrene cup. The solution has an initial temperature of 21.0 °C. He quickly adds an excess of zinc powder and records the maximum temperature as 57.5 °C. Assume that the solution has a specific heat capacity of 4.18 J g^{-1} K^{-1} and a density of 1.00 g cm^{-3}.
   a. Calculate the heat energy change in this experiment. *(1 mark)*
   b. Calculate the enthalpy change of reaction in kJ mol^{-1}, for:
   $$Zn(s) + CuSO_4(aq) \rightarrow Cu(s) + ZnSO_4(aq)$$
   (3 marks)

3. Combustion of 1.50 g of propan-1-ol, C_3H_7OH, raised the temperature of 150 cm^3 of water by 64.0 °C.
   Calculate the enthalpy change of combustion of propan-1-ol to an appropriate number of significant figures. *(4 marks)*

9.3 Bond enthalpies

Specification reference: 3.2.1

Breaking and making bonds

Energy has to be put in to break bonds (endothermic) while energy is released when bonds are formed (exothermic).

During chemical reactions there is usually a difference between the energy involved in bond breaking and bonds making.

- When more energy is released in forming new bonds than is needed to break bonds, the overall energy change is exothermic (ΔH –ve).
- When more energy is needed to break bonds than is released when new bonds are formed, the overall energy change is endothermic (ΔH +ve).

Activation energy, E_a

The energy required to break bonds acts as an energy barrier to a reaction, known as the activation energy (E_a). Activation energy is the minimum energy required for a reaction to take place.

Figures 1 and 2 show enthalpy profile diagrams for exothermic and endothermic reactions and the role of activation energy, E_a.

Activation energy is an important factor in determining whether a reaction will take place.

- Reactions with small activation energies generally take place very rapidly, because the energy needed to break bonds is readily available from the surroundings.
- Very large activation energies may present such a large energy barrier that a reaction may take place extremely slowly or even not at all.

Average bond enthalpy

The strength of different bonds varies and is measured as bond enthalpy. The stronger a bond is, the harder the bond is to break, and the more endothermic the bond enthalpy.

The average bond enthalpy is the mean amount of energy required to break one mole of a specified type of covalent bond in a gaseous molecules

$$A–B(g) \rightarrow A(g) + B(g)$$

▼ **Table 1** *Examples of average bond enthalpies*

Bond	Average bond enthalpy / kJ mol^{-1}
H–H	+436
C–C	+347
C–H	+413
C=C	+612

A hydrogen $H_2(g)$, molecule contains 1 H–H bond:

- 436 kJ of energy is needed to break 1 mol of H–H bonds
- 436 kJ of energy is released when 1 mol of H–H bonds are formed.

An ethane $C_2H_6(g)$ molecule contains 6 C–H bonds and 1 C–C bond:

$6 \times 413 + 1 \times 347 = 2825$ kJ of energy is involved in breaking and making all the bonds in 1 mol of $C_2H_6(g)$.

Synoptic link

See Topic 9.1, Enthalpy changes, for more details of exothermic and endothermic reactions and enthalpy profile diagrams.

Synoptic link

You will learn more about activation energy and its relevance to reaction rates in Chapter 10, Reaction rates and equilibrium.

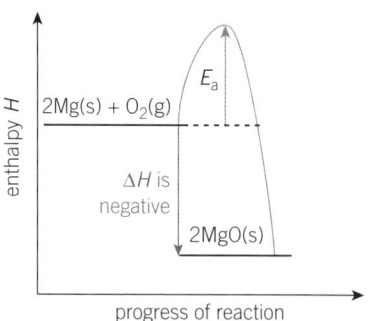

▲ **Figure 1** *Exothermic enthalpy profile diagram*

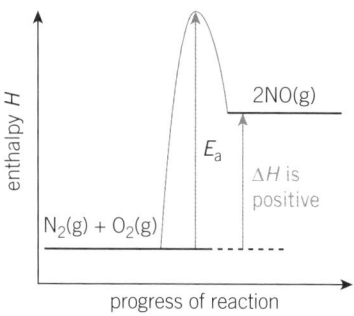

▲ **Figure 2** *Endothermic enthalpy profile diagram*

Synoptic link

See Topic 5.3, Covalent bonding, where bond enthalpy was first introduced.

Enthalpy

Synoptic link
Determination of enthalpy changes from experiments is covered in Topic 9.2, Measuring enthalpy changes. Hess' Law is covered in Topic 9.4, Hess' Law.

Limitations of using average bond enthalpies
The average bond enthalpy for a bond is an average value obtained from many molecules containing the bond. The actual value for a particular bond in a molecule may be slightly different from the average value. This means that calculations involving average bond enthalpies may give different enthalpy values than those obtained from experiments and other methods such as using Hess' law.

Using average bond enthalpies in calculations
Average bond enthalpies can be used to work out the enthalpy change of some reactions.

For a reaction involving gaseous molecules:

$$\Delta_r H = \Sigma(\text{bond enthalpies in reactants}) - \Sigma(\text{bond enthalpies in products})$$

Revision tip
Σ = 'sum of'

Revision tip
It is always easiest to draw out displayed formulae so that you can count the number of bonds that are broken and made.

> **Worked example: Calculating an enthalpy change from bond enthalpies**
>
> Calculate the enthalpy change for the combustion of 1 mol of methane, $CH_4(g)$ (see Table 2 for average bond enthalpies):
>
> $$CH_4(g) + 2O_2(g) \rightarrow CO_2(g) + 2H_2O(g)$$
>
> **Step 1:** Calculate the energy required to break the bonds in the reactants:
>
> $4 \times C-H = 4 \times 413 = 1652 \text{ kJ mol}^{-1}$
>
> $2 \times O=O = 2 \times 498 = 996 \text{ kJ mol}^{-1}$
>
> Total = 2648 kJ mol^{-1}
>
> **Step 2:** Calculate the energy released when bonds in products are formed:
>
> $2 \times C=O = 2 \times 805 = 1610 \text{ kJ mol}^{-1}$
>
> $4 \times O-H = 4 \times 464 = 1856 \text{ kJ mol}^{-1}$
>
> Total = 3466 kJ mol^{-1}
>
> **Step 3:** Calculate the enthalpy change:
>
> $\Delta H = \Sigma(\text{bond enthalpies in reactants}) - \Sigma(\text{bond enthalpies in products})$
>
> $= 2648 - 3466 = -818 \text{ kJ mol}^{-1}$

▼ **Table 2** *Average bond enthalpies*

Bond	Average bond enthalpy / kJ mol^{-1}
C–H	+413
O=O	+498
C=O	+805
O–H	+464
C=C	+612

Summary questions

1. Define the term 'average bond enthalpy'. *(2 marks)*

2. Why does the actual value of bond enthalpy sometimes differ from the average bond enthalpy for the bond? *(1 mark)*

3. Calculate the enthalpy change for the complete combustion of 1 mol of ethene, C_2H_4 (see Table 2 for average bond enthalpies). *(3 marks)*

9.4 Hess' law and enthalpy cycles

Specification reference: 3.2.1

Hess' law

Hess' law is used to calculate enthalpy changes that are not easy to measure directly in experiments.

Hess' law states that, if a reaction can take place by more than one route, and the starting and finishing conditions are the same, the total enthalpy change is the same for each route.

Using Hess' law with enthalpy changes of formation, $\Delta_f H^\ominus$

You can work out the standard enthalpy change of any reaction from the standard enthalpy changes of formation ($\Delta_f H^\ominus$) of the reactants and products.

The energy cycle in Figure 1 shows how the enthalpy change of a reaction ($\Delta_r H$) is linked to the enthalpy changes of formation of the reactants and products. The elements link the reactants and products so that $\Delta_f H$ values can be used.

> **Synoptic link**
>
> See Topic 9.1, Enthalpy changes, for more details of enthalpy changes of formation.

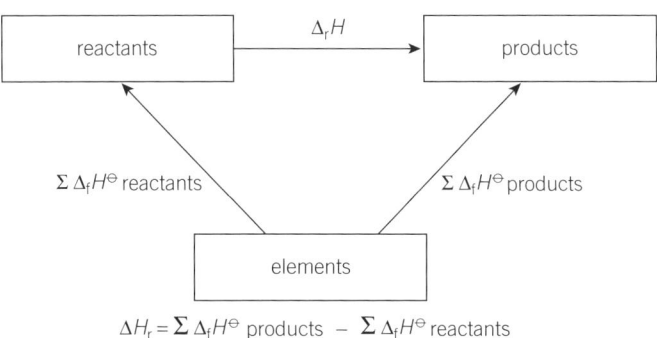

▲ **Figure 1** *Energy cycle using $\Delta_f H$*

> **Worked example: Calculating an enthalpy change from enthalpy changes of formation, $\Delta_f H^\ominus$**
>
> Calculate the enthalpy change for the reaction below:
>
> $$C_2H_4(g) + 3O_2(g) \rightarrow 2CO_2(g) + 2H_2O(l)$$
>
> Standard enthalpy changes of formation are shown in Table 1.
>
> The enthalpy cycle linking reactants and products is shown below, together with the enthalpy changes for the formation of reactants and products from elements.
>
>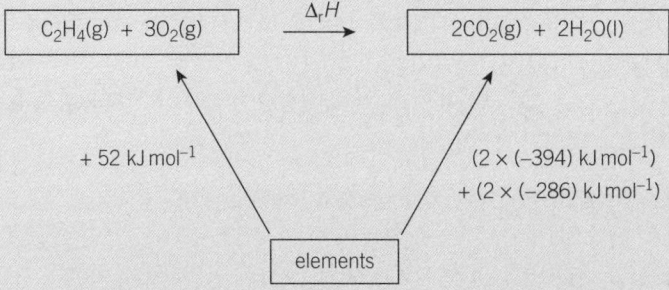
>
> ▲ **Figure 2** *The Hess' law diagram for the combustion of ethene*
>
> $\Delta_r H = \sum \Delta_f H^\ominus \text{ products} - \sum \Delta_f H^\ominus \text{ reactants}$
>
> $\therefore \Delta_r H = [(2 \times -394) + (2 \times -286)] - (+52) = -1412 \text{ kJ mol}^{-1}$

▼ **Table 1** *Standard enthalpy changes of formation*

Substance	$\Delta_f H^\ominus$ / kJ mol^{-1}
$C_2H_4(g)$	+52
$CO_2(g)$	−394
$H_2O(l)$	−286

> **Revision tip**
>
> The enthalpy change of formation of oxygen has not been listed. But remember that $\Delta_f H$ for an element is zero.

Enthalpy

> **Synoptic link**
> See Topic 9.1, Enthalpy changes, for more details of enthalpy changes of combustion.

Using Hess' law with enthalpy changes of combustion, $\Delta_c H^\ominus$

You can work out a standard enthalpy change of some reactions from the standard enthalpy changes of combustion ($\Delta_c H^\ominus$) of the reactants and products.

The energy cycle in Figure 3 shows how the enthalpy change of a reaction ($\Delta_r H$) is linked to the enthalpy changes of combustion of the reactants and products. The combustion products link the reactants and products so that $\Delta_c H$ values can be used.

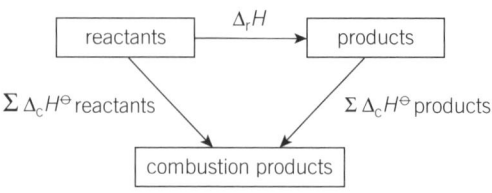

$\Delta_r H = \Sigma \Delta_c H^\ominus \text{ reactants} - \Sigma \Delta_c H^\ominus \text{ products}$

▲ **Figure 3** *Energy cycle using $\Delta_c H$*

> **Revision tip**
> Compare the direction of the arrows in the energy cycles for formation and combustion. They go in different directions.
>
> Also notice that the enthalpy changes swap over in the calculation:
>
> Formation: products – reactants
>
> Combustion: reactants – products

Worked example: Calculating an enthalpy change from enthalpy changes of combustion, $\Delta_c H^\ominus$

Calculate the enthalpy change for the reaction below:

$$C(s) + 2H_2(g) \rightarrow CH_4(g)$$

Standard enthalpy changes of combustion are shown in Table 1.

The enthalpy cycle linking reactants and products is shown below, together with the enthalpy changes for the combustion of reactants and products.

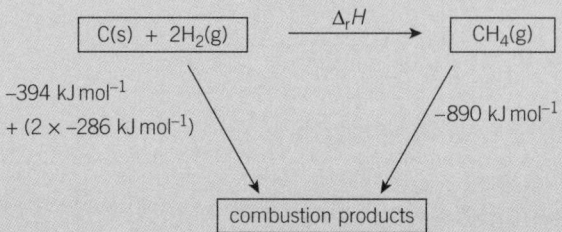

▲ **Figure 4** *The Hess' law diagram for the formation of methane*

$\Delta_r H = \sum \Delta_c H^\ominus \text{ reactants} - \sum \Delta_c H^\ominus \text{ products}$

$\therefore \Delta_r H = [(-394) + (2 \times -286)] - (-890) = -76 \text{ kJ mol}^{-1}$

▼ **Table 2** *Standard enthalpy changes of combustion*

Substance	$\Delta_c H^\ominus$ / kJ mol^{-1}
C(s)	−394
H_2(g)	−286
CH_4(g)	−890

Summary questions

1. Using the $\Delta_f H$ information below, calculate $\Delta_r H$ for the equation:
   $$CH_4(g) + H_2O(g) \rightarrow CO(g) + 3H_2(g)$$
   (*2 marks*)
   $\Delta_f H(CH_4(g)) = -75 \text{ kJ mol}^{-1}$; $\Delta_f H(H_2O(g)) = -242 \text{ kJ mol}^{-1}$;
   $\Delta_f H(CO(g)) = -110 \text{ kJ mol}^{-1}$

2. Use the $\Delta_c H$ data below to calculate $\Delta_f H$ of pentane, C_5H_{12}:
   $$5C(s) + 6H_2(g) \rightarrow C_5H_{12}(l)$$
   (*3 marks*)
   $\Delta_c H(C(s)) = -393.5 \text{ kJ mol}^{-1}$; $\Delta_c H(H_2(g)) = -285.8 \text{ kJ mol}^{-1}$;
   $\Delta_c H(C_5H_{12}(l)) = -3509.1 \text{ kJ mol}^{-1}$

3. Calculate the enthalpy change for the decomposition of magnesium nitrate shown below:
   $$2Mg(NO_3)_2(s) \rightarrow 2MgO(s) + 4NO_2(g) + O_2(g)$$
   (*3 marks*)
   $\Delta_f H(Mg(NO_3)_2) = -791 \text{ kJ mol}^{-1}$; $\Delta_f H(MgO) = -602 \text{ kJ mol}^{-1}$;
   $\Delta_f H(NO_2) = -33 \text{ kJ mol}^{-1}$

Chapter 9 Practice questions

1. 50.0 cm³ of solution **A** is mixed with 50.0 cm³ of solution **B**. The temperature increased from 19.5 °C to 35.0 °C. The specific heat capacity and density of the mixture is the same as for water.

   How much energy, in kJ, is released?

   A 3.240
   B 6.480
   C 3240
   D 6480 (*1 mark*)

2. Dinitrogen oxide, N_2O, reacts as follows

   $$2N_2O(g) \rightarrow 2N_2(g) + O_2(g) \qquad \Delta_r H = -164 \text{ kJ mol}^{-1}$$

   What is the value for the enthalpy change of formation of $N_2O(g)$?

   A −164
   B −82
   C +82
   D +164 (*1 mark*)

3. Combustion of a 0.9525 g sample of carbon disulfide, $CS_2(l)$, increases the temperature of 150 cm³ of water from 22.5 °C to 34.0 °C.

   a Write the equation, with state symbols, for the complete combustion of carbon disulfide under standard conditions. (*2 marks*)

   b Determine the enthalpy change of combustion of $CS_2(l)$. Give your answer to **three** significant figures. (*4 marks*)

   c The experimental enthalpy change of combustion of $CS_2(l)$ is much less exothermic than the actual value from data books.

   Suggest **two** reasons for this difference. (*2 marks*)

4. Enthalpy changes can be calculated indirectly using enthalpy changes of formation and bond enthalpies.

   a The equation that represents the enthalpy change of combustion of pentane, C_5H_{12}, is shown below.

   $$C_5H_{12}(l) + 8O_2(g) \rightarrow 5CO_2(g) + 6H_2O(l)$$

   This enthalpy change can be found using the enthalpy changes of formation, $\Delta_f H$, in the table.

Compound	$C_5H_{12}(l)$	$O_2(g)$	$CO_2(g)$	$H_2O(l)$
$\Delta_f H$ / kJ mol⁻¹	−173.2	0	−393.5	−285.8

   i Define the term *standard enthalpy change of formation*.
   Include the standard conditions in your answer. (*3 marks*)

   ii Explain the value for $\Delta_f H$ of $O_2(g)$. (*1 mark*)

   iii Calculate the enthalpy change of combustion of pentane. (*3 marks*)

   b The reaction between hydrazine, N_2H_4, and hydrogen peroxide, H_2O_2, has been used to propel rockets. The equation for the reaction under these conditions is shown below.

   $$N_2H_4(g) + 2H_2O_2(g) \rightarrow N_2(g) + 4H_2O(g)$$
   $$\Delta_r H = -787 \text{ kJ mol}^{-1}$$

   The bond enthalpy for the N≡N bond in $N_2(g)$ can be calculated using the enthalpy change above and the bond enthalpies shown in the table.

bond	N—H	N—N	O—H	O—O
bond enthalpy / kJ mol⁻¹	+391	+158	+464	+144

   i What is meant by the term *bond enthalpy*? (*2 marks*)

   ii Use the bond enthalpies above to calculate the bond enthalpy for the N≡N bond. (*3 marks*)

10.1 Reaction rates

Specification reference: 2.1.3, 3.2.2

The rate of a chemical reaction

The rate of a chemical reaction is the change in concentration of a product or reactant over time:

$$\text{rate} = \frac{\text{change in concentration}}{\text{time}}$$

The rate of a reaction can be found by measuring how a reactant is used up, or a product is made, over time. The method chosen will depend on the physical properties of the reactant or products.

- If a reaction produces a gas, the rate of reaction could be monitored by measuring the increase in gas volume over time, or the decrease in the mass of the reactants.
- If a reactant is an acid, you could measure the increase in pH over time to monitor the rate of reaction as the acid is used up.

Collision theory

You can use collision theory to understand how conditions affect the rate of a chemical reaction.

For a reaction to occur, particles must collide.

- The colliding particles must have enough energy to break the existing bonds.
- The particles must be in the correct orientation so that the reactive parts of the particles collide.

As a result, only a small proportion of collisions between particles result in a reaction.

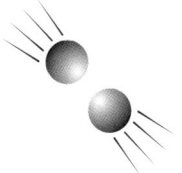

▲ Figure 1 *Particles must collide before they can react*

- Activation energy, E_a, is the minimum energy for a reaction to take place.
- Different reactions have different activation energies.
- The lower the activation energy, the larger the number of particles that can react and the faster the reaction rate.

Synoptic link

Activation energy was introduced in Topic 9.3, Bond enthalpies.

For more details, see Topic 10.2, Catalysts, and Topic 10.3, The Boltzmann distribution.

Key term

Activation energy, E_a: The minimum energy for a reaction to take place.

The effect of concentration on the rate of reaction

Changing the concentration affects the rate of reaction for reactions that involve solutions and gases.

Concentration and reactions involving solutions

When the concentration of a solution is increased, there are more particles in the same volume:

- the reactant particles are closer together
- collisions are more frequent
- the rate increases.

Pressure and reactions involving gases

When the pressure of a gas is increased, there is the same number of gas molecules in a smaller volume, increasing the concentration. The molecules collide more frequently and the rate increases.

Reaction rates and equilibrium

Rates of reaction from graphs

Concentration–time graphs

The rate of reaction can be calculated from a concentration–time graph. In Figure 2, the curve shows how the concentration of a reactant decreases over time.

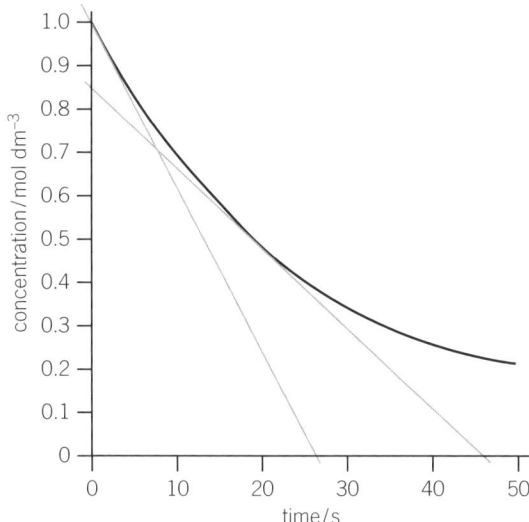

▲ **Figure 2** *A concentration–time graph*

To find the rate at any time:

- draw a tangent to the curve at that time
- measure the gradient of the tangent.

The gradient gives the rate of the reaction at that time.

Two tangents are shown on Figure 2:

At $t = 0$ s, rate $= \dfrac{1.0 - 0.0}{27 - 0} = \dfrac{1.0}{27} = 0.037 \text{ mol dm}^{-3}\text{ s}^{-1}$

The rate at $t = 0$ is called the initial rate of reaction.

At $t = 20$ s, rate $= \dfrac{0.85 - 0}{46 - 0} = \dfrac{0.85}{46} = 0.018 \text{ mol dm}^{-3}\text{ s}^{-1}$

As the reaction proceeds, the reactants are used up, concentration decreases, and the rate of reaction decreases.

Using other graphs to determine rates

For some reactions, it may be easier to measure the formation of a gas over time, or to measure the decrease in mass over time.

The same method is used to find a reaction rate – a tangent is drawn at a given time and the gradient gives the rate.

> **Revision tip**
>
> The gradient of a tangent measures how steep the tangent is:
>
> Gradient $= \dfrac{\text{difference in } y}{\text{difference in } x}$
>
> The steeper the line, the greater the gradient and the faster the rate.
>
> Rate is concentration/time.
>
> The units of rate are: mol dm^{-3} s^{-1}.

Summary questions

1 What is the activation energy of a reaction? *(1 mark)*

2 Explain how increasing the pressure affects the rate of a reaction involving gases. *(2 marks)*

3 a Draw a concentration–time graph from the experimental results. *(2 marks)*

Time / s	0	10	20	30	40	50	60	70	80	90	100
Concentration of product / mol dm^{-3}	0.00	0.34	0.50	0.64	0.75	0.85	0.91	0.95	0.98	1.00	1.00

   b Use your graph to calculate the initial rate of reaction. *(2 marks)*

10.2 Catalysts
Specification reference: 3.2.2

Catalysts
- A catalyst increases the rate of a reaction without being used up over the whole reaction.
- A catalyst allows a reaction to proceed via a different route with a lower activation energy.

In the presence of a catalyst, more particles have sufficient energy to react at the lower activation energy, increasing the rate of reaction. The enthalpy profile diagrams in Figures 1 and 2 show the reduction in activation energy in the presence of a catalyst for exothermic and endothermic reactions.

Synoptic links
You learnt about activation energies for endothermic and exothermic reactions in Topic 9.3, Bond enthalpies.

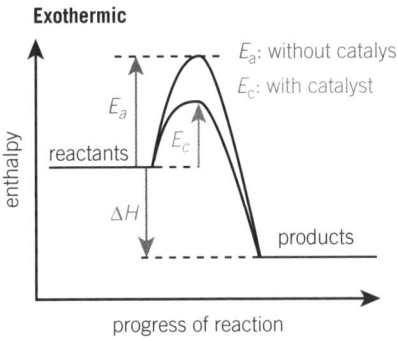

▲ **Figure 1** *Exothermic reaction, with and without a catalyst*

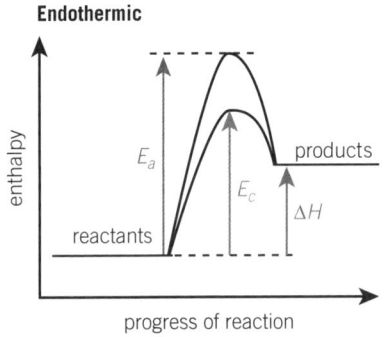

▲ **Figure 2** *Endothermic reaction, with and without a catalyst*

Homogeneous and heterogeneous catalysis
Homogeneous catalyst
A homogeneous catalyst is in the same state as the reactants.

In the upper atmosphere, ozone, O_3, breaks down to form oxygen molecules, O_2:

$$O_3(g) + O(g) \rightarrow 2O_2(g)$$

This reaction is catalysed by chlorine radicals, $Cl\bullet(g)$, produced from the breakdown of CFC molecules. A chlorine radical is a homogeneous catalyst because a $Cl\bullet(g)$ radical is a gas and has the same physical state as the reactants $O_3(g)$ and $O(g)$, which are also gases.

Heterogeneous catalyst
A heterogeneous catalyst is in a different state from the reactants.

The Haber process is used to produce ammonia from nitrogen and hydrogen gases:

$$N_2(g) + 3H_2(g) \rightleftharpoons 2NH_3(g)$$

The reaction is catalysed by solid iron. Iron is a heterogeneous catalyst because iron is a solid and has a different physical state from the reactants, $N_2(g)$ and $H_2(g)$, which are gases.

Synoptic links
You will learn more about the catalytic breakdown of ozone in Topic 15.2, Organohalogen compounds in the environment.

Economic and environmental benefits of using catalysts

Many reactions used in industrial processes require high temperatures and pressures, which are expensive to generate.

Industry often uses catalysts to lower the temperature and pressure required for reactions to take place. Lower temperatures reduce the energy demand, less fossil fuels are burnt, and less carbon dioxide gas is produced. This increases the sustainability of an industrial process and reduces the emission of CO_2 as a greenhouse gas. Reduced energy demand also has considerable cost savings.

Summary questions

1. Explain why catalysts increase the rate of chemical reactions. *(2 marks)*

2. What are the economic and environmental benefits of using catalysts? *(2 marks)*

3. Explain, with examples, what is meant by the terms:
   a. homogeneous catalyst *(2 marks)*
   b. heterogeneous catalyst. *(2 marks)*

10.3 The Boltzmann distribution
Specification reference: 3.2.2

A Boltzmann distribution is used to represent the energy of the particles in a sample of a gas at a given temperature. See Figure 1.

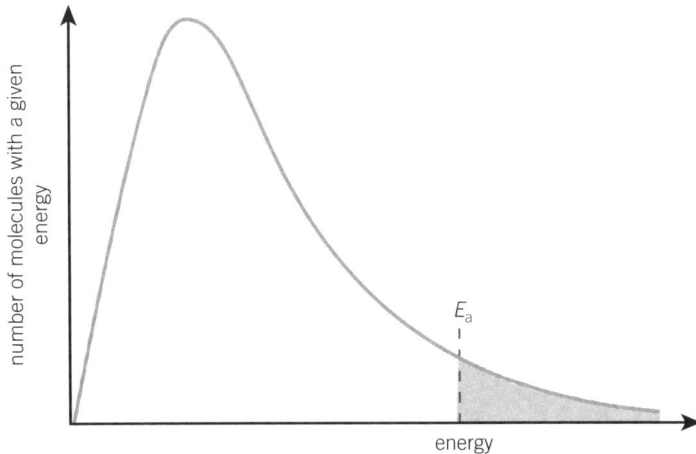

▲ **Figure 1** *The Boltzman distribution of molecular energies*

- The distribution curve is not symmetrical.
- Most of the particles have an energy, which falls within a comparatively narrow range, with fewer particles having much more or much less energy.
- The line does not cross the *x*-axis at higher energy. It would only do so at infinite energy.
- The line starts at the origin, showing that no particles have zero energy.
- The total area under the distribution curve is equal to the total number of gas particles.

Activation energy and the Boltzmann distribution

In Figure 1, the shaded part of the distribution curve shows the proportion of particles that exceed the activation energy, E_a, of the reaction.

The larger the shaded area:

- the greater the number of particles that have an energy greater than E_a
- the faster the reaction rate.

Boltzmann distributions and temperature

Figure 2 shows the Boltzmann distribution for a gas sample at two different temperatures T_1 and T_2, where T_2 is a higher temperature than T_1.

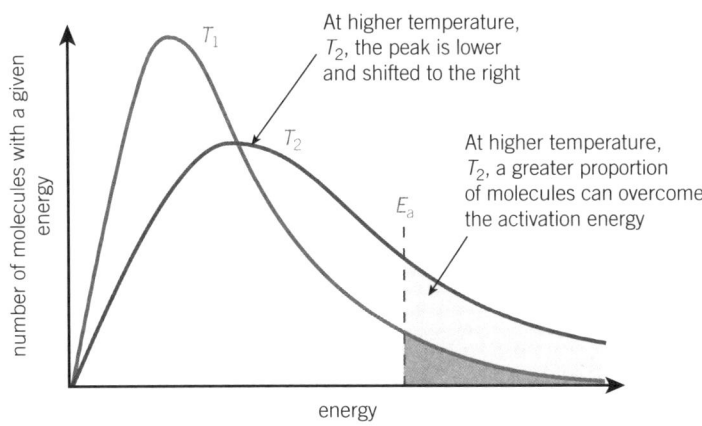

▲ **Figure 2** *The effect of temperature on the Boltzman distribution*

Revision tip
Activation energy is the minimum energy for a reaction to take place. For a reaction to take place, colliding particles must have energy equal to or greater than the activation energy.

Synoptic link
For more details on activation energy, see Topic 9.3, Bond enthalpies, Topic 10.1, Reaction rates, and Topic 10.2, Catalysts.

Reaction rates and equilibrium

> **Revision tip**
> As the temperature increases, a much higher proportion of particles have energy greater than or equal to the activation energy, E_a.
>
> Even a small increase in temperature can lead to a very large increase in the rate of a chemical reaction. An increase of 10 °C doubles the rate of many reactions.

As the temperature increases:

- the average energy of the particles increases
- the peak of the distribution curve moves to the right
- the distribution curve becomes flatter because the total number of particles remains the same.
- a greater proportion of molecules exceeds the activation energy and more molecules are able to react (see the shaded areas in Figure 2)
- the rate of reaction increases.

Catalysts and Boltzmann distributions

Adding a catalyst does not change the distribution curve, but the activation energy is now at a lower energy (E_c in Figure 3).

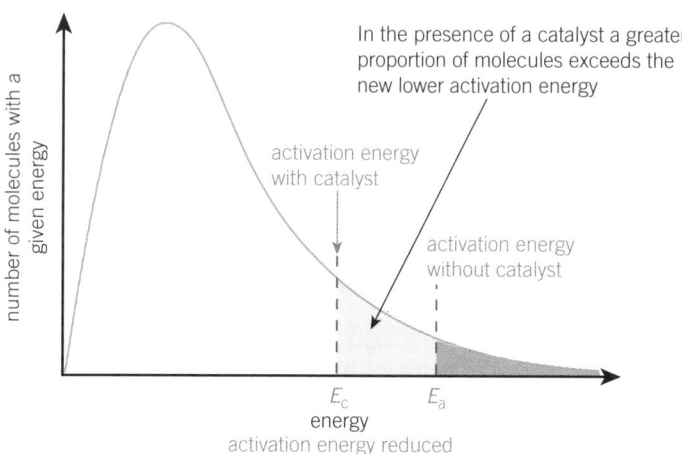

▲ **Figure 3** *The effect of a catalyst on the number of molecules with enough energy to react*

> **Synoptic links**
> See Topic 10.2, Catalysts, for the effect of catalysts on the activation energy in enthalpy profile diagrams.

Although the shape of the distribution curve is identical:

- there is a greater proportion of molecules that exceed the new lower activation energy, E_c (see the shaded areas in Figure 3)
- on collision, more molecules will react to form products
- the rate of reaction increases.

Summary questions

1. In a Boltzmann distribution what does the total area under the curve represent? *(1 mark)*

2. Use the Boltzmann distribution to explain how a catalyst changes the rate of reaction. *(3 marks)*

3. Use the Boltzmann distribution to explain how the rate would change when the temperature is decreased by 10 °C. *(3 marks)*

10.4 Dynamic equilibrium and le Chatelier's principle

Specification reference: 3.2.3

Reversible reactions and dynamic equilibria

Many reactions are reversible; they can go in the forward and reverse directions. For example:

$$2SO_2(g) + O_2(g) \rightleftharpoons 2SO_3(g)$$

The $\rightleftharpoons$ sign indicates that the reversible reaction is in equilibrium (see below).

Dynamic equilibrium

A closed system is isolated from the surroundings – nothing can enter and nothing can leave.

If a reversible reaction takes place in a 'closed system', a dynamic equilibrium can be set up where both the forward and reverse reactions are taking place.

At equilibrium:

- as fast as the reactants are reacting to form products, the products are reacting to form reactants
- the concentrations of the reactants and products remain unchanged
- the equilibrium position is somewhere between reactants and products.

> **Synoptic link**
>
> See Topic 9.1, Enthalpy changes, for details of the chemical system and the surroundings.

Le Chatelier's principle

Le Chatelier's principle allows us to predict how changes in conditions may affect the position of equilibrium.

The effect of changing concentration on the equilibrium position

The equilibrium position shifts to minimise a change in concentration.

- When the concentration of one of the reactants is increased, the equilibrium position shifts to the right to decrease the concentration.
- When the concentration of one of the reactants is decreased, the equilibrium position shifts to the left to increase the concentration.

> **Key term**
>
> **Le Chatelier's principle:** When a system in dynamic equilibrium is subjected to a change in conditions, the equilibrium position shifts to minimise the change.

The effect of changing temperature on the equilibrium position

The equilibrium position shifts to minimise a change in temperature.

- If the forward reaction is exothermic (ΔH –ve) the reverse reaction is endothermic (ΔH +ve).
- When the temperature is increased, the position of equilibrium shifts in the endothermic direction to decrease the temperature.
- When the temperature is decreased, the position of equilibrium shifts in the exothermic direction to increase the temperature.

> **Revision tip**
>
> Concentration changes only affect equilibria involving solutions or gases.

> **Worked example: The effect of temperature on equilibrium position**
>
> Industrially, SO_2 and O_2 are reacted to produce SO_3. The forward reaction is exothermic.
>
> ΔH +197 kJ mol^{-1} $2SO_2(g) + O_2(g) \rightleftharpoons 2SO_3(g)$ ΔH –197 kJ mol^{-1}
> Endothermic   Exothermic
>
> - When the temperature is increased, the position of equilibrium shifts in the endothermic direction (ΔH = +197 kJ mol^{-1}) to the left, reducing the yield of SO_3.
> - When the temperature is decreased, the position of equilibrium shifts in the exothermic direction (ΔH = –197 kJ mol^{-1}) to the right, increasing the yield of SO_3.

> **Revision tip**
>
> **Rules for temperature**
>
> An increase in temperature shifts the equilibrium position in the endothermic direction.
>
> A decrease in temperature shifts the equilibrium position in the exothermic direction.

Reaction rates and equilibrium

> **Revision tip**
> Changes in pressure only affect equilibria involving gases.

> **Revision tip**
> **Rules for pressure:**
> An increase in pressure shifts the equilibrium position to the side with fewer gas molecules.
>
> A decrease in pressure shifts the equilibrium position to the side with more gas molecules.

The effect of changing pressure on the equilibrium position

The equilibrium position shifts to minimise a change in pressure.

- When the pressure is increased, the equilibrium position shifts to the side with fewer gas molecules to decrease the pressure.
- When the pressure is decreased, the equilibrium position shifts to the side with more gas molecules to increase the pressure.

If an equilibrium has the same number of gas molecules on both sides of the reaction, changing the pressure has no effect on the position of equilibrium.

Worked example: The effect of pressure on equilibrium position

The equilibrium between $SO_2(g)$, $O_2(g)$, and $SO_3(g)$ is shown below.

$$2SO_2(g) + O_2(g) \rightleftharpoons 2SO_3(g)$$
$$3 \text{ mol} \rightleftharpoons 2 \text{ mol}$$

- When the pressure is increased, the equilibrium position shifts towards fewer gas molecules (3 mol → 2 mol), to the right, increasing the yield of SO_3.
- When the pressure is decreased, the equilibrium position shifts towards more gas molecules (3 mol ← 2 mol), to the left, decreasing the yield of SO_3.

The effect of catalysts on the position of equilibrium

A catalyst increases the rate of a chemical reaction but is not itself used up.

- A catalyst increases the rate of the forward reaction by the same amount as the rate of reverse reaction is increased.
- A catalyst does not affect the position of equilibrium. It simply allows equilibrium to be reached in less time.

The importance of compromise for industrial processes

The chemical industry uses reversible reactions to produces many important chemicals. For example, the production of ammonia gas:

$$N_2(g) + 3H_2(g) \rightleftharpoons 2NH_3(g) \qquad \Delta H = -92 \text{ kJ mol}^{-1}$$

Le Chatelier's principle can be used to predict the conditions of temperature and pressure needed to obtain a maximum equilibrium yield of ammonia:

- Low temperature: The forward reaction is exothermic.
- High pressure: There are fewer gas molecules on the right.

However, there are operational difficulties in using these conditions.

- A low temperature may produce a reaction rate so slow that equilibrium may not be reached.
- A high pressure requires a large quantity of energy, increasing the cost of the process. High pressures also present safety concerns from potential toxic gas leaks, endangering the workforce and the environment.

In practice, conditions are used to obtain a reasonable equilibrium yield without compromising the rate of reaction, excessive costs, and risks to safety.

> **Summary questions**
>
> 1 What does the sign $\rightleftharpoons$ mean? *(1 mark)*
>
> 2 Predict, with reasons, how increasing the temperature of an endothermic reaction affects the rate of reaction and position of equilibrium. *(2 marks)*
>
> 3 Methanol, CH_3OH, can be made by industry using a reversible reaction:
> $$CO(g) + 2H_2(g) \rightleftharpoons CH_3OH(g)$$
> $\Delta H = -76 \text{ kJ mol}^{-1}$
>
> a Predict, with reasons, the conditions of temperature and pressure for a maximum equilibrium yield of methanol. *(2 marks)*
>
> b Explain why the actual operational conditions used may be different. *(2 marks)*

10.5 The equilibrium constant, K_c

Specification reference: 3.2.3

The equilibrium law

The position of equilibrium can be worked out from the stoichiometric equation for a reversible reaction using the equilibrium law.

For the equilibrium: $aA + bB \rightleftharpoons cC + dD \qquad K_c = \dfrac{[C]^c [D]^d}{[A]^a [B]^b}$

- K_c is the equilibrium constant in terms of concentration.
- The square brackets indicate the concentration of each species, in $mol\,dm^{-3}$.

> **Revision tip**
> Remember $K_c = \dfrac{[products]}{[reactants]}$

Calculating an equilibrium constant, K_c

For an equilibrium equation, the equilibrium constant, K_c, can be calculated from the equilibrium concentrations of reactants and products.

> 🖩 **Worked example: Calculating K_c from equilibrium concentrations**
>
> $N_2(g)$, $H_2(g)$, and $NH_3(g)$ are present in the equilibrium below. The equilibrium concentrations are also given.
>
> $N_2(g) + 3H_2(g) \rightleftharpoons 2NH_3(g)$
>
> $N_2(g) = 2.13\,mol\,dm^{-3}$; $H_2(g) = 1.38\,mol\,dm^{-3}$; $NH_3(g) = 0.0825\,mol\,dm^{-3}$
>
> Calculate the numerical value for the equilibrium constant, K_c.
>
> Give your answer to an appropriate number of significant figures.
>
> $K_c = \dfrac{[NH_3(g)]^2}{[N_2(g)][H_2(g)]^3} = \dfrac{0.0825^2}{2.13 \times 1.38^3} = 1.215\,880\,895 \times 10^{-3}$
>
> $= 1.22 \times 10^{-3}$ to three significant figures

> **Revision tip**
> Make sure you give your answer to an appropriate number of significant figures.
>
> You should round just **once**, for the final value using the least number of significant figures in the data provided.
>
> In this example, all concentrations are to three significant figures so your final answer should also be to three significant figures.

The magnitude of the equilibrium constant, K_c

The numerical value of K_c gives a rough guide to the equilibrium position.

- $K_c = 1$: equilibrium position is halfway between reactants and products.
- $K_c > 1$: equilibrium position favours the right-hand or products side.
- $K_c < 1$: equilibrium position favours the left-hand or reactants side.

> **Summary questions**
>
> 1. Write the expression for K_c for the equilibrium below.
> $2SO_2(g) + O_2(g) \rightleftharpoons 2SO_3(g)$ *(1 mark)*
>
> 2. $H_2(g)$, $I_2(g)$, and $HI(g)$ are present in the equilibrium below.
> $H_2(g) + I_2(g) \rightleftharpoons 2HI(g)$
> $H_2(g) = 0.0520\,mol\,dm^{-3}$; $I_2(g) = 0.125\,mol\,dm^{-3}$; $HI(g) = 0.321\,mol\,dm^{-3}$
>   Calculate the value of K_c to **three** significant figures. *(3 marks)*
>
> 3. $CO(g)$, $H_2(g)$, and $CH_3OH(g)$ are present in the equilibrium below.
> $CO(g) + 2H_2(g) \rightleftharpoons CH_3OH(g)$
> $CO(g) = 0.310\,mol\,dm^{-3}$; $H_2(g) = 0.240\,mol\,dm^{-3}$; Value of $K_c = 14.6$
>   Calculate the equilibrium concentration of CH_3OH to an appropriate number of significant figures. *(3 marks)*

Chapter 10 Practice questions

1. Which statement is correct when a suitable catalyst is used?
   A. The Boltzmann distribution shifts to the right.
   B. The area under the curve increases.
   C. The activation energy shifts to lower energy.
   D. The mean energy of the molecules increases. *(1 mark)*

2. In industry, sulfuric acid is made from sulfur trioxide. The equilibrium below shows the reaction for the formation of sulfur trioxide.
   $$2SO_2(g) + O_2(g) \rightleftharpoons 2SO_3(g) \quad \Delta H = -196 \text{ kJ mol}^{-1}$$
   Which change increases the equilibrium yield of SO_3 the most?
   A. Increasing pressure and increasing temperature.
   B. Decreasing pressure and increasing temperature.
   C. Increasing pressure and decreasing temperature.
   D. Decreasing pressure and decreasing temperature. *(1 mark)*

3. A chemist set up the equilibrium below.
   $$2N_2O_4(g) + O_2(g) \rightleftharpoons 2N_2O_5(g)$$
   The equilibrium concentrations, in $mol\,dm^{-3}$, are:
   $[N_2O_4(g)]$ 0.150, $[O_2(g)]$ 0.150, $[N_2O_5(g)]$ 2.50.
   What is the numerical value of K_c to three significant figures?
   A. 5.40×10^{-4} B. 111 C. 2.40×10^{-2} D. 1850 *(1 mark)*

4. Reaction rates can be increased or decreased by changing conditions of pressure or temperature, or by carrying out the reaction in the presence of a suitable catalyst.
   a. Describe and explain the effect of increasing the pressure on the rate of a reaction. *(2 marks)*
   b. Explain how increasing the temperature increases the rate of reaction.
   Include a labelled sketch of the Boltzmann distribution. Label the axes. *(4 marks)*
   c. Using a labelled enthalpy profile diagram, show how a catalyst increases the rate of a chemical reaction with a negative ΔH value. *(3 marks)*

5. A chemist set up the following equilibrium:
   $$2NO(g) + O_2(g) \rightleftharpoons 2NO_2(g) \quad \Delta H = -115 \text{ kJ mol}^{-1}$$
   a. The system reached a dynamic equilibrium.
   State **two** features of a *dynamic equilibrium*. *(2 marks)*
   b. The temperature of the equilibrium mixture is increased.
   Explain what happens to the position of the equilibrium. *(2 marks)*
   c. The pressure of the equilibrium mixture is increased.
   Explain what happens to the position of the equilibrium. *(2 marks)*
   d. Under these conditions, $K_c = 1.57 \times 10^5 \text{ dm}^3 \text{mol}^{-1}$.
   The equilibrium concentrations of $NO(g)$ and $O_2(g)$ are shown below.
   $[NO(g)]$ $3.0 \times 10^{-3} \text{ mol dm}^{-3}$ $[O_2(g)]$ $1.5 \times 10^{-3} \text{ mol dm}^{-3}$
   i. Write the expression for K_c for this equilibrium. *(1 mark)*
   ii. Calculate the equilibrium concentration of $NO_2(g)$.
   Give your answer to **two** significant figures and in standard form. *(2 marks)*

11.1 Organic chemistry
Specification reference: 4.1.1

Hydrocarbons

Saturated and unsaturated hydrocarbons

Hydrocarbons contain carbon and hydrogen atoms only.

A **saturated hydrocarbon** (e.g. butane) contains single carbon–carbon bonds only.

An **unsaturated hydrocarbon** (e.g. ethene) contains at least one multiple carbon–carbon bond, including C=C, C≡C, or aromatic rings.

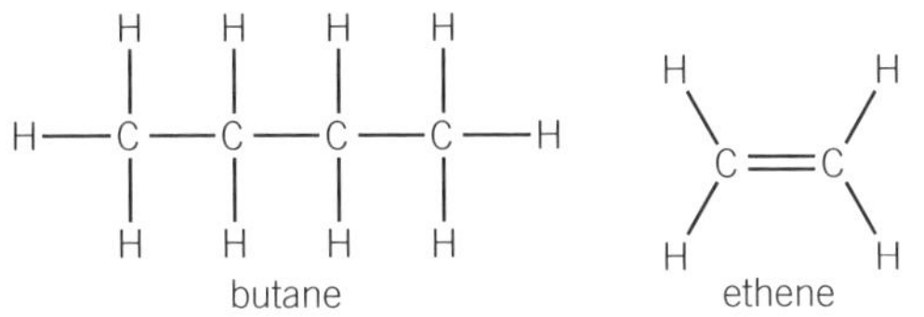

▲ **Figure 1** *Butane (saturated) and ethene (unsaturated)*

Aliphatic hydrocarbons

Aliphatic hydrocarbons contain carbon atoms joined together in straight chains, branched chains, or non-aromatic rings.

A **straight-chain hydrocarbon** contains one continuous carbon chain only.

A **branched-chain hydrocarbon** contains a shorter side chain bonded to a longer continuous carbon chain.

An **alicyclic hydrocarbon** contains carbon atoms joined together in non-aromatic rings with or without side chains.

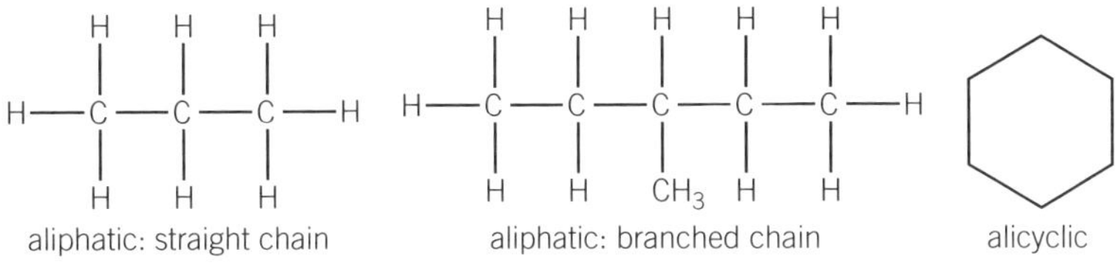

▲ **Figure 2** *Different types of aliphatic hydrocarbon*

Aromatic hydrocarbons

An aromatic hydrocarbon (e.g. benzene) is an unsaturated hydrocarbon that contains a benzene ring.

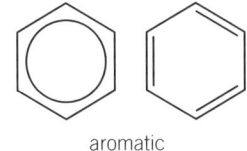

▲ **Figure 3** *Two ways of showing the structure of benzene, C_6H_6*

> **Revision tip**
> Aliphatic hydrocarbons can be saturated (single C–C bonds only), or unsaturated (contain at least one C=C, C≡C bond).

> **Synoptic link**
> Alicyclic and aromatic rings are usually shown as skeletal formulae. For details, see Topic 11.3, Representing the formulae of organic compounds.

> **Revision tip**
> Aromatic hydrocarbons are unsaturated hydrocarbons that contain a benzene ring.

> **Synoptic link**
> Aromatic rings are covered in detail in Topic 11.2, Nomenclature of organic compounds.

Basic concepts of organic chemistry

▼ Table 1 *The alkanes homologous series*

Name	Molecular formula
Methane	CH_4
Ethane	C_2H_6
Propane	C_3H_8
Butane	C_4H_{10}

Key term

Homologous series: A series of organic compounds having the same functional group but with each successive member differing by CH_2.

Key term

Functional group: The group of atoms responsible for the characteristic reactions of a compound.

Revision tip

Saturated alicyclic hydrocarbons have the general formula C_nH_{2n}.

Synoptic link

For more details of molecular formulae, see Topic 3.2, Determination of formulae.

Homologous series and functional groups

Homologous series

A **homologous series** is a series of organic compounds that has the same functional group. Each successive member of a homologous series differs from the previous member by the addition of CH_2. Table 1 shows the first four members of the alkanes homologous series. Note that the formula of each successive member differs by CH_2.

Functional group

A **functional group** is a group of atoms responsible for the characteristic reactions of a compound.

The **general formula** is the simplest algebraic formula of a member of a homologous series.

Table 2 shows the functional group and general formula of the four homologous series that are studied during the first year of the course.

▼ Table 2 *The functional group and general formula of different homologous series*

Homologous series	Functional group	General formula
Alkanes	C—C	C_nH_{2n+2}
Alkenes	C=C	C_nH_{2n}
Alcohols	C—OH	$C_nH_{2n+1}OH$
Haloalkanes	C—X	$C_nH_{2n+1}X$

X = halogen atom: F, Cl, Br, or I

Summary questions

1 What is meant by the following terms?
   a saturated hydrocarbon
   b aromatic hydrocarbon
   c homologous series *(4 marks)*

2 What is the molecular formula for the following?
   a The alkane with 7 carbon atoms. *(1 mark)*
   b The alkene with 25 carbon atoms. *(1 mark)*
   c The chloroalkane with 6 carbon atoms. *(1 mark)*

3 Octyne, C_8H_{14} is a member of the alkynes homologous series.
   a What is the general formula for the alkynes? *(1 mark)*
   b What is the molecular formula for the alkyne with 12 carbon atoms? *(1 mark)*

11.2 Nomenclature of organic compounds

Specification reference: 4.1.1

Nomenclature

Nomenclature is the system of naming organic compounds.

Carbon chains

The **stem** of a name indicates the number of carbon atoms in the longest continuous carbon chain.

A shorter carbon side chain is called an **alkyl group**.

- An alkyl group has the general formula C_nH_{2n+1}, with one fewer H atom than the parent alkane group that gives the name.
- R is often used for any alkyl group in a homologous series, e.g. ROH indicates any alcohol.

▼ **Table 1** Naming carbon chains

Number of carbon atoms	Name of stem	Alkane	Alkyl group
1	meth-	methane, CH_4	methyl, CH_3
2	eth-	ethane, C_2H_6	ethyl, C_2H_5
3	prop-	propane, C_3H_8	propyl, C_3H_7
4	but-	butane, C_4H_{10}	butyl, C_4H_9
5	pent-	pentane, C_5H_{12}	pentyl, C_5H_{11}
6	hex-	hexane, C_6H_{14}	hexyl, C_6H_{13}
7	hept-	heptane, C_7H_{16}	heptyl, C_7H_{15}
8	oct-	octane, C_8H_{18}	octyl, C_8H_{17}
9	non-	nonane, C_9H_{20}	nonyl, C_9H_{19}
10	dec-	decane, $C_{10}H_{22}$	decyl, $C_{10}H_{21}$

> **Revision tip**
>
> You need to learn the names of the first 10 members of the alkanes homologous group and their corresponding alkyl groups.
>
> You should also be able to write their formulae.
>
> Alkanes: C_nH_{2n+2}
>
> Alkyl group: C_nH_{2n+1}

Naming organic compounds

A **prefix** indicates the presence and position of alkyl groups or functional groups.

A **suffix** indicates the presence and position of functional groups.

▼ **Table 2** Naming organic compounds

Homologous series	Functional group	Prefix	Suffix
Alkane	C—C		-ane
Alkene	C=C		-ene
Haloalkane	C—Cl	chloro-	
	C—Br	bromo-	
	C—I	iodo-	
Alcohol	C—OH	hydroxy-	-ol

Basic concepts of organic chemistry

Positioning alkyl groups and functional groups

Alkyl groups or functional groups can be bonded to different C atoms in the longest carbon chain. A number indicates the position, counting from the end of the longest carbon chain that gives the smallest number.

A prefix is added to show the number of identical groups:

    di- for 2 tri- for 3 tetra- for 4

Ordering alkyl groups and functional groups

If there is more than one alkyl group or functional group, they are ordered alphabetically.

Examples of names for organic structures

Figure 1 shows examples of naming organic compounds. Notice the use of hyphens and commas.

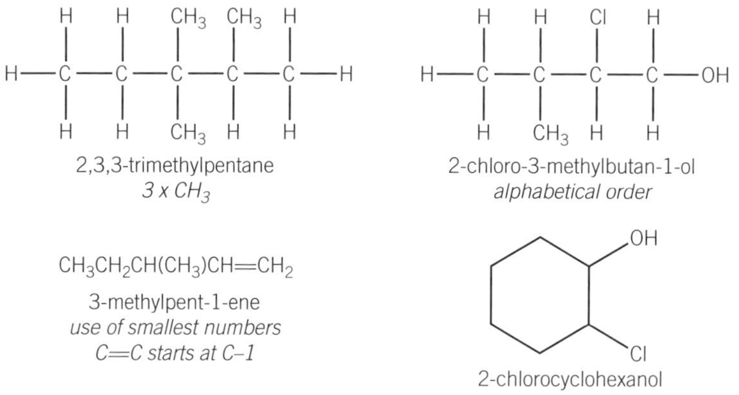

2,3,3-trimethylpentane
3 × CH_3

2-chloro-3-methylbutan-1-ol
alphabetical order

$CH_3CH_2CH(CH_3)CH=CH_2$
3-methylpent-1-ene
use of smallest numbers
C=C starts at C–1

2-chlorocyclohexanol

▲ **Figure 1** *Naming organic compounds*

Summary questions

1 a What is meant by an 'alkyl' group, R? *(1 mark)*
 b What is the molecular formula of:
  i nonane **ii** a heptyl group? *(2 marks)*

2 Name the following compounds:
 a $CH_3CH_2CH_2OH$
 b $CH_3CHBrCH_2CH_3$
 c $CH_3CH=CHCH_2CH_3$
 d $CH_3CH_2CH(CH_3)CH(CH_3)_2$
 e $CH_3CH_2CH_2CH_2CH(CH_3)CH_2Br$ *(5 marks)*

3 Draw the structures for the following:
 a 2,2,4-trimethylpentane
 b 3,4-dimethylhex-2-ene
 c 3-methylcyclohexanol *(3 marks)*

11.3 Representing the formulae of organic compounds

Specification reference: 4.1.1

Organic chemists use several different types of formula for organic compounds, which are described below.

Figure 1 shows the differences between each formula for 2-methylpentane.

Empirical formula
The **empirical formula** is the simplest whole number ratio of atoms of each element present in a compound.

Molecular formula
The **molecular formula** is the actual number and type of atoms of each element in a molecule of a compound.

Structural formula
A **structural formula** is the minimal detail that shows the arrangement of atoms in a molecule.

Displayed formula
A **displayed formula** shows the relative positioning of atoms and the bonds between them.

Skeletal formula
A **skeletal formula** is a simplified organic formula shown by removing hydrogen atoms from alkyl chains, leaving just a carbon skeleton and associated functional groups.

Synoptic link
For more details of molecular and empirical formulae, see Topic 3.2, Determination of formulae.

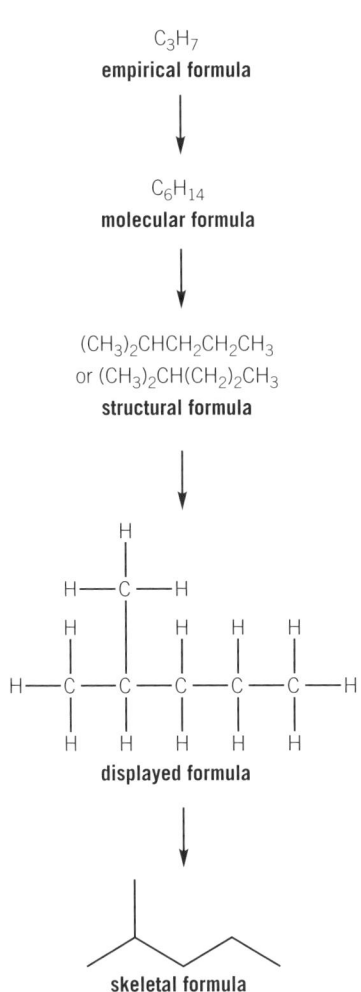

▲ **Figure 1** *Different types of formula for 2-methylpentane*

Summary questions

1. But-1-ene is an alkene.
   What are the molecular, empirical, structural, displayed, and skeletal formulae of but-1-ene? *(5 marks)*

2. The displayed formula of erythritol, a naturally occurring sugar, is shown below:

   HO—CH₂—CH(OH)—CH(OH)—CH₂—OH (displayed)

   What are the skeletal, structural, molecular, and empirical formulae of erythritol? *(4 marks)*

3. The skeletal formula of limonene, the smell in oranges, is shown below:

   What are the molecular and empirical formulae of limonene? *(2 marks)*

11.4 Isomerism

Specification reference: 4.1.1

Isomerism

Isomerism occurs when two or more organic molecules have the same molecular formula but different arrangements of atoms.

Structural isomers

Structural isomers are compounds with the same molecular formula but different structural formulae.

> **Revision tip**
> When drawing structures of alcohols, the bond to the OH group must always go to the O of the OH group. See how the bonds to the three OH groups in propane-1,2,3-triol should be shown:
>
> HO—C—C—C—OH with H, OH, H below and H, H, H above

> **Worked example 1: Identifying structural isomers of haloalkanes**
>
> What are the structural isomers of C_3H_7Br?
>
> The bromine atom can be bonded to C1 at the end of the carbon chain, or to C2 in the middle of the carbon chain. This gives two structural isomers.
>
> 1-bromopropane   2-bromopropane 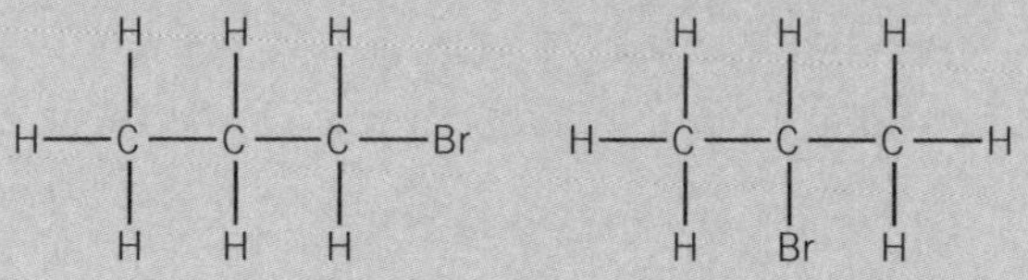
>
> ▲ Figure 1 Structural isomers of C_3H_7Br

> **Worked example 2: Identifying structural isomers of alkenes**
>
> What are the structural isomers of C_4H_8 that are alkenes?
>
> The carbon skeleton can have four carbons in a straight chain or three carbons in a straight chain and a one-carbon side chain attached to C2:
>
> - The straight-chain isomer can have the C=C double bond between C1 and C2 or between C2 and C3.
> - The branched-chain isomer can only have the C=C double bond between C1 and C2.
> - This gives three structural isomers, shown here as skeletal formulae: 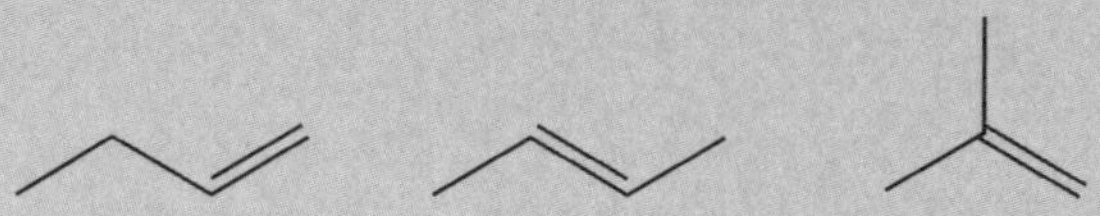
>
> but-1-ene   but-2-ene   methylpropene
>
> ▲ Figure 2 Structural isomers of C_4H_8 that are alkenes

> **Revision tip**
> Notice that we only need one number in the name to show the position of a C=C bond.
>
> e.g. C=C between C1 and C2 is '-1-ene'.
>
> In methylpropene, there is only one possible position for the methyl group and so a number is not needed.

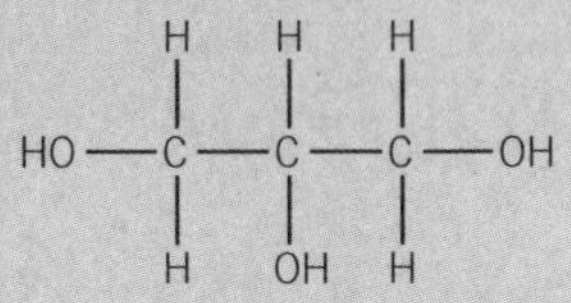

Summary questions

1. What are structural isomers? *(1 mark)*

2. Draw and name the four structural isomers of C_4H_9Cl. *(8 marks)*

3. Draw skeletal formulae and name the structural isomers of C_5H_{10} that are alkenes. *(5 marks)*

11.5 Introduction to reaction mechanisms
Specification reference: 4.1.1

Covalent bond fission
Covalent bonds can be broken by homolytic fission or heterolytic fission.

Homolytic fission
In homolytic fission, a covalent bond is broken with each bonding atom receiving one electron from the bonded pair to form two radicals. A radical is a species with an unpaired electron, shown by a dot (•) e.g.:

$$Cl–Cl \rightarrow Cl• + •Cl$$

Heterolytic fission
In heterolytic fission, a covalent bond is broken with one of the bonded atoms receives both electrons from the bonded pair, forming two oppositely charged ions, e.g.:

$$H–Cl \rightarrow H^+ + :Cl^-$$

Reaction mechanisms and curly arrows
Reaction mechanisms show how a reaction takes place.

- Curly arrows are used to show the movement of a pair of electrons.
- The direction of the arrow shows the movement of the electron pair when bonds are being broken or made (see Figure 1).
- A curly arrow must start from a bond, a lone pair, or a negative charge.
- When a bond is broken, the curly arrow shows heterolytic fission.

Types of reaction
Addition reactions
In an addition reaction, two reactant molecules join together to form a single product. For example, H_2O (as steam) adds across the C=C double bond in but-2-ene to form butan-2-ol.

$$CH_3CH=CHCH_3 + H_2O \rightarrow CH_3CH_2CHOHCH_3$$

Substitution reactions
In a substitution reaction, an atom or group of atoms exchanges with a different atom or group of atoms. For example, an OH^- ion exchanges with a Br^- ion in 1-bromopropane to form propan-1-ol.

$$CH_3CH_2CH_2Br + OH^- \rightarrow CH_3CH_2CH_2OH + Br^-$$

Elimination reactions
In an elimination reaction, a molecule loses a small molecule, such as water. For example, H_2O is eliminated from propan-1-ol to form propene and water.

$$CH_3CH_2CH_2OH \rightarrow CH_3CH=CH_2 + H_2O$$

Key term
Radical: A very reactive species with an unpaired electron.

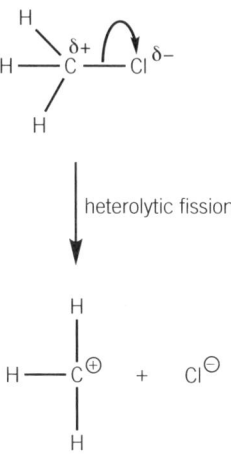

▲ **Figure 1** *Curly arrow during heterolytic bond fission*

Revision tip
Notice the difference:

Addition: 2 molecules → 1 molecule

Elimination: 1 molecule → 2 molecules

Summary questions
1. What is a radical? *(1 mark)*
2. What does a curly arrow in a reaction mechanism indicate? *(1 mark)*
3. Consider the equation below.
   $$C_2H_4 + H_2 \rightarrow C_2H_6$$
   What type of reaction is taking place? *(1 mark)*

Chapter 11 Practice questions

1 What is the general formula of an alkyl group?
   A C_nH_{2n-1}
   B C_nH_{2n}
   C C_nH_{2n+1}
   D C_nH_{2n+2} *(1 mark)*

2 What is the empirical formula for the compound shown?

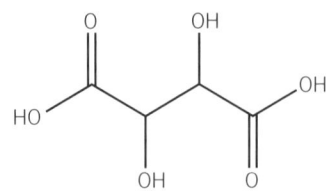

   A $C_2H_2O_3$
   B $C_2H_3O_3$
   C $C_4H_4O_6$
   D $C_4H_6O_6$ *(1 mark)*

3 How many structural isomers does C_5H_{12} have?
   A 2
   B 3
   C 4
   D 5 *(1 mark)*

4 What is the systematic name for the compound shown?
   A 2-ethylpent-3-ene
   B 4-ethylpent-2-ene
   C 3-methylhex-4-ene
   D 4-methylhex-2-ene *(1 mark)*

5 This question is about terms used in organic chemistry and the compounds **A–F**.

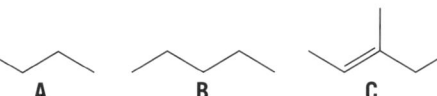
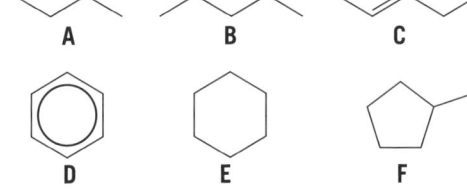

   a i What is meant by a *saturated hydrocarbon*? *(2 marks)*
     ii Which structures are saturated hydrocarbons? *(1 mark)*
   b What is meant by the terms *homologous series* and *functional group*? *(3 marks)*
   c i Which structures are aliphatic? *(1 mark)*
     ii Which structures are alicyclic? *(1 mark)*
   d Which structure has the empirical formula CH? *(1 mark)*
   e i What is meant by *structural isomers*? *(1 mark)*
     ii Which compounds are structural isomers of one another? *(1 mark)*
   f Three organic reactions are shown below.
     For each equation,
     • classify the type of reaction
     • state the functional group in the organic compound formed.
     i $C_2H_4 + H_2O \rightarrow CH_3CH_2OH$ *(2 marks)*
     ii $C_2H_6 + Br_2 \rightarrow C_2H_5Br + HBr$ *(2 marks)*
     iii $CH_3CH_2CH_2OH \rightarrow CH_3CH=CH_2 + H_2O$ *(2 marks)*

6 Compound **G** has the percentage composition by mass: C, 85.71%; H, 14.29%. The relative molecular mass of compound **G** is 56.0.
   a Calculate the molecular formula of compound **G**. Show your working. *(3 marks)*
   b Compound **G** is one of four structural isomers. Three isomers are aliphatic with no ring, and one is alicyclic.
     i Draw the skeletal formulae of the four structural isomers.
     ii Write the systematic name for each isomer. *(4 marks)*

12.1 Properties of the alkanes

Specification reference: 4.1.2

Alkanes

Alkanes have the general formula C_nH_{2n+2}.

The alkanes are saturated hydrocarbons. This means that alkanes have single covalent bonds and contain carbon and hydrogen only. The C–C and C–H bonds in alkane molecules are σ-bonds (sigma bonds):

- Sigma bonds are formed from the overlap of orbitals directly between the bonding atoms.
- Sigma bonds can freely rotate.

Shapes of alkanes

In alkane molecules, each carbon atom is surrounded by four bonded pairs of electrons. The four bonded pairs repel each other equally, resulting in a tetrahedral shape around each carbon atom, and bond angles of 109.5°.

The properties of alkanes

Boiling points

Straight-chain alkanes

As the carbon chain length increases, boiling points increase (see Table 1):

- There are more electrons and more points of contact between molecules.
- The strength of the London forces between the molecules increases.
- More energy is needed to break the intermolecular forces.

Branched-chain alkanes

As branching in isomers of alkanes increases, boiling points decrease (see Table 2):

- There are fewer points of contact between the molecules.
- The strength of the London forces between the molecules decreases.
- Less energy is needed to break the intermolecular forces.

Solubility

Alkanes are non-polar molecules and do not dissolve in polar solvents like water. However, they will mix with non-polar solvents.

Synoptic link

For names and formulae of alkanes, see Topic 11.2, Nomenclature of organic compounds.

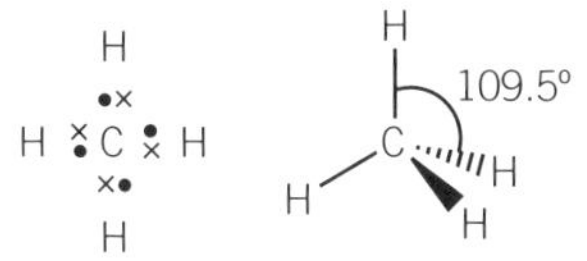

▲ Figure 1 *The tetrahedral shape of methane*

Synoptic link

For further details about the shapes of molecules, see Topic 6.1, Shapes of molecules and ions.

▼ Table 1 *Effect of chain length on boiling point*

Alkane	Boiling point /°C
$CH_3CH_2CH_3$	−42
$CH_3CH_2CH_2CH_3$	−1
$CH_3CH_2CH_2CH_2CH_3$	36

▼ Table 2 *Effect of branching on boiling point of C_5H_{12} isomers*

Alkane	Boiling point /°C
$CH_3CH_2CH_2CH_2CH_3$	36
$(CH_3)_2CHCH_2CH_3$	28
$(CH_3)_2C(CH_3)_2$	9

Synoptic link

You learned about London forces and the factors affecting their size in Topic 6.3, Intermolecular forces.

Summary questions

1. How is a sigma bond formed? *(1 mark)*

2. Describe and explain the shape and bond angles around the carbon atoms in alkanes. *(4 marks)*

3. State and explain the trend in boiling points of straight-chain alkanes. *(3 marks)*

12.2 Chemical reactions of the alkanes

Specification reference: 4.1.2

> **Synoptic link**
>
> For details on electronegativity, see Topic 6.2, Electronegativity and polarity, and, for bond enthalpies, Topic 9.3, Bond enthalpies.

Reactivity

Alkanes have a low reactivity:

- The C–C and C–H sigma bonds are very strong and hard to break. These bonds have a high bond enthalpy.
- The C–C and C–H bonds are non-polar. There is little difference in electronegativity between C and H atoms.

Fuels and combustion

Fuels are substances that can be burnt in oxygen to release heat energy. Although alkanes are relatively unreactive, many alkanes react exothermically with oxygen and make good fuels.

Complete combustion

Alkanes burn completely in an unlimited supply of oxygen to produce carbon dioxide and water in an exothermic reaction.

For methane, CH_4, (the main alkane in natural gas):

$$CH_4(g) + 2O_2(g) \rightarrow CO_2(g) + 2H_2O(l)$$

For octane, C_8H_{18}, (the main alkane in petrol), the products are identical:

$$C_8H_{18}(l) + 12\tfrac{1}{2}O_2(g) \rightarrow 8CO_2(g) + 9H_2O(l)$$

Incomplete combustion

In a limited supply of oxygen, incomplete combustion occurs.

- The hydrogen in the hydrocarbon still forms water.
- The carbon only undergoes partial oxidation to form carbon monoxide, CO.
- Unburnt carbon particles can also be released as carbon particles (soot).

Equations for the incomplete combustion of methane and octane are shown below:

$$CH_4(g) + 1\tfrac{1}{2}O_2(g) \rightarrow CO(g) + 2H_2O(l)$$

$$C_8H_{18}(l) + 8\tfrac{1}{2}O_2(g) \rightarrow 8CO(g) + 9H_2O(l)$$

- Carbon monoxide is a toxic gas that binds to haemoglobin in red blood cells, preventing haemoglobin from carrying oxygen around the body.
- Carbon monoxide is difficult to detect as it is colourless and odourless.

Reaction of alkanes with halogens

Methane reacts with chlorine in the presence of ultraviolet (UV) radiation by radical substitution. The overall equation is shown below:

$$CH_4 + Cl_2 \rightarrow CH_3Cl + HCl$$

Radical substitution mechanism

The mechanism proceeds by three steps: initiation, propagation, and termination.

> **Key term**
>
> **Radical:** A species with an unpaired electron.

Alkanes

Initiation

UV radiation provides energy for homolytic fission of chlorine molecules, producing highly reactive chlorine radicals (Cl•).

$$Cl_2 \rightarrow 2Cl\bullet$$

Propagation

Propagation proceeds by two reactions. This step produces the HCl and CH_3Cl products in the overall equation.

1 Chlorine radicals react with methane to form HCl and methyl radicals ($CH_3\bullet$):

$$Cl\bullet + CH_4 \rightarrow HCl + \bullet CH_3$$

2 Methyl radicals react with chlorine to form CH_3Cl and a chlorine radical:

$$\bullet CH_3 + Cl_2 \rightarrow CH_3Cl + Cl\bullet$$

In Reaction 1, Cl• reacts and $CH_3\bullet$ forms.

In Reaction 2, $CH_3\bullet$ reacts and Cl• forms.

- Radicals are being consumed and then regenerated.
- Propagation can carry on as a chain reaction, provided that there is a constant supply of the reactants CH_4 and Cl_2 and a supply of UV.

If you add together the equations for the two propagation reactions, you get the overall equation for the reaction.

Termination

In each termination step, two radicals react together to form a new molecule. This happens naturally if two radicals collide and react. The reaction then stops as there is no longer a constant supply of radicals. There are three possible termination reactions involving Cl• and $\bullet CH_3$ radicals:

1 $Cl\bullet + Cl\bullet \rightarrow Cl_2$
2 $\bullet CH_3 + Cl\bullet \rightarrow CH_3Cl$
3 $\bullet CH_3 + \bullet CH_3 \rightarrow C_2H_6$

Limitations of radical substitution

Radical substitution has limitations for organic synthesis because a large mixture of products can be formed.

Further substitution

The CH_3Cl product still contains hydrogen atoms. These hydrogen atoms can be replaced by further substitution with chlorine to produce a mixture of other products:

$$CH_3Cl + Cl_2 \rightarrow CH_2Cl_2 + HCl$$
$$CH_2Cl_2 + Cl_2 \rightarrow CHCl_3 + HCl$$
$$CHCl_3 + Cl_2 \rightarrow CCl_4 + HCl$$

Formation of structural isomers

For alkanes with a chain of three or more carbon atoms, substitution can take place at different positions along the carbon chain, giving a mixture of structural isomers. For example, radical substitution of propane with chlorine could form 1-chloropropane and 2-chloropropane:

$$CH_3CH_2CH_3 + Cl_2 \rightarrow CH_3CH_2CH_2Cl + HCl$$
1-chloropropane

$$CH_3CH_2CH_3 + Cl_2 \rightarrow CH_3CHClCH_3 + HCl$$
2-chloropropane

> **Synoptic link**
>
> See Topic 11.5, Introduction to reaction mechanisms, for an introduction into mechanisms and homolytic fission.

> **Revision tip**
>
> There are three steps in the radical substitution of methane: initiation, propagation, and termination.
>
> - In the initiation step, radicals are made.
> - In the propagation steps, radicals react with molecules to form new molecules and radicals.
> - In the termination steps, two radicals join together. A mixture of new molecules is formed.

> **Summary questions**
>
> 1 Alkanes reacts with halogens.
>   a State the essential conditions. *(1 mark)*
>   b Name the mechanism. *(1 mark)*
>
> 2 Write equations for the complete and incomplete combustion of butane. *(2 marks)*
>
> 3 Ethane can react with chlorine to produce chloroethane.
>   a Write equations for:
>   i the initiation step *(1 mark)*
>   ii the propagation steps *(2 marks)*
>   iii all possible termination steps. *(3 marks)*
>   b Write the structural formulae of all the possible organic products that could be formed by further substitution of chloroethane. *(3 marks)*

Chapter 12 Practice questions

1 Which equation shows a reaction that takes place during incomplete combustion of ethane?

　A $C_2H_6 + 3\frac{1}{2}O_2 \rightarrow 2CO_2 + 3H_2O$　　B $C_2H_6 + 2\frac{1}{2}O_2 \rightarrow 2CO + 3H_2O$

　C $C_2H_6 + 2O_2 \rightarrow 2CO_2 + 3H_2$　　D $C_2H_6 + O_2 \rightarrow 2CO + 3H_2$　　(1 mark)

2 Which alkane has the lowest boiling point?

　A $CH_3CH_2CH_2CH_2CH_3$　　B $(CH_3)_2CHCH_2CH_3$

　C $CH_3CH_2CH_2CH_2CH_2CH_3$　　D $CH_3C(CH_3)_2CH_3$　　(1 mark)

3 Bromine reacts with methylbutane in UV.

A propagation reaction forms the radical X•

$$CH_3CH(CH_3)CH_2CH_3 + Br• \rightarrow X• + HBr$$

How many structural formulae of X• are possible?

　A 2　　B 3　　C 4　　D 5　　(1 mark)

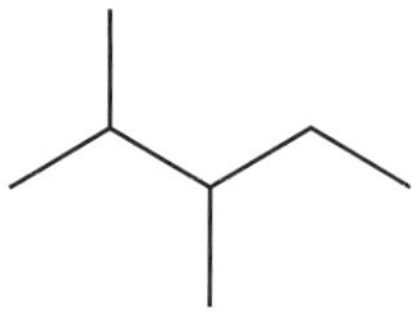

4 An alkane has the skeletal formula in Figure 1.

What is the structural formula of the alkane?

　A $CH_3CH_2CH_2CH(CH_3)CH_2CH_3$　　B $(CH_3)_2CH_2CH_2CH_2CH_2CH_3$

　C $(CH_3)_2CHCH_2CH(CH_3)CH_3$　　D $(CH_3)_2CHCH(CH_3)CH_2CH_3$　(1 mark)

▲ Figure 1

5 A student attempts to prepare 1-bromopropane by reacting propane with bromine in the presence of UV.

　a Write an equation for this reaction.　(1 mark)

　b i Name the mechanism.　(1 mark)

　　ii Explain why homolytic fission takes place.　(1 mark)

　　iii Outline the mechanism for this reaction. Show propane as C_3H_8.
    Include all possible termination steps.　(7 marks)

　c The reaction produces only a low yield of 1-bromopropane. The product was contaminated with termination products and other organic impurities.

　　Suggest **two** reasons for the formation of the other organic impurities.

　　For each reason, show the structural formula of a possible impurity.

　　　(4 marks)

6 Crude oil comprises a mixture of hydrocarbons including many alkanes.

　a What is the molecular formula of the alkane with 32 carbon atoms?

　　　(1 mark)

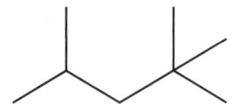

▲ Figure 2

　b Write the systematic name of the alkane in Figure 2　(1 mark)

　c The boiling points of alkanes are shown in Table 1.

　　Describe and explain the trend shown by the boiling points of the hydrocarbons in the table.　(6 marks)

▼ Table 1

Alkane	Boiling point / K
propane	231
butane	273
pentane	309
2-methylbutane	301
2,2-dimethylpropane	283

　d Petrol contains a mixture of alkanes.

In parts **i** and **ii**, petrol is assumed to consist solely of octane.

　　i Write the equation, including state symbols, for the complete combustion of octane.　(2 marks)

　　ii A car travels 1 km and produces 72.0 dm³ of CO_2, measured at room temperature and pressure (RTP). Calculate the mass of octane that has been burnt and the volume of air required at RTP. Assume that complete combustion has taken place and that air contains 20% of oxygen by volume.　(5 marks)

13.1 The properties of alkenes
Specification reference: 4.1.3

Introducing alkenes
Alkenes are unsaturated hydrocarbons containing the C=C functional group. Alkene molecules can have more than one double bond and can have straight chains, branched chains, or rings. Straight chain or branched-chain alkenes with one C=C double bond have the general formula C_nH_{2n}.

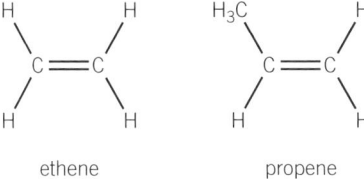

▲ **Figure 1** Ethene, C_2H_4, and propene, C_3H_6, the first two members of the alkenes homologous series

The C=C double bond
The C=C double bond consists of a π-bond (pi bond) and a σ-bond (sigma bond):

- The π-bond (pi bond) is the sideways overlap of adjacent p orbitals above and below the bonding C atoms. The π-bond prevents rotation about the C=C bond.
- The σ-bond (sigma bond) is the overlap of orbitals directly between the bonding atoms.

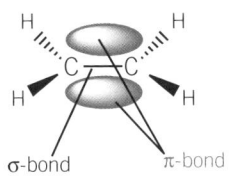

▲ **Figure 2** The σ-bond and π-bond in a C=C double bond

The shape around the C=C bond
There are three regions of electron density around each carbon atom in the C=C bond (one double bond and two single bonds). The three regions of electron density repel each other as far apart as possible to give a trigonal planar shape with a bond angle of 120°.

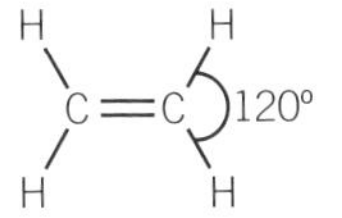

▲ **Figure 3** The trigonal planar shape about the C atoms of the C=C bond in an ethene molecule

Synoptic link
You met sigma bonds (σ-bonds) in Topic 12.1, Properties of the alkanes.

Synoptic link
For further details about the shapes of molecules, see Topic 6.1, Shapes of molecules and ions.

Reactions of alkenes
Alkenes are much more reactive than alkanes. The reactivity of alkenes is caused by the presence of a π-bond:

- The π-bond introduces a region of high electron density above and below the plane of the bonding C atoms in the C=C bond. Electron-deficient species called **electrophiles** can attack the high electron density of the π-bond, causing the alkene to react.
- In the C=C bond, the π-bond has a smaller bond enthalpy than the σ-bond and is broken more easily.

Synoptic link
For further details about electrophiles, see Topic 13.4, Electrophilic addition in alkenes.

Revision tip:
Alkanes are saturated hydrocarbons and contain single C–C bonds only, which are σ-bonds.

Alkenes are unsaturated hydrocarbons and contain at least one C=C bond, composed of a σ-bond and a π-bond.

Key terms
σ-bond: A bond caused by orbital overlap directly between the bonding atoms.

π-bond: A bond caused by sideways overlap of adjacent p orbitals above and below the bonding atoms.

Unsaturated: An unsaturated hydrocarbon contains at least one multiple bond.

Summary questions

1. How is a π-bond formed? *(1 mark)*

2. Explain the shape and bond angle around the C atoms of a C=C bond. *(4 marks)*

3. Why are alkenes much more reactive than alkanes? *(3 marks)*

13.2 Stereoisomerism

Specification reference: 4.1.3

> **Key term**
>
> **Stereoisomers:** Have the same structural formula but a different arrangement of the atoms in space.

> **Synoptic link**
>
> Optical isomerism is studied in the second year of the A level course. It does not form part of AS Chemistry.

Stereoisomers

Alkenes have structural isomers, but the C=C double bond allows some alkenes to also have stereoisomers.

Stereoisomers have the same structural formula but a different arrangement of the atoms in space.

There are two types of stereoisomerism:

- E/Z isomerism which occurs in some alkenes.
- Optical isomerism which occurs in a range of different compounds.

E/Z stereoisomerism

The π-bond restricts rotation about the C=C double bond. The groups attached to the carbon atoms of the C=C bond have fixed positions and cannot rotate from one side of the double bond to the other side.

A molecule has E/Z isomers if it meets the following conditions:

- There is a C=C double bond.
- There are two different groups attached to each carbon atom of the C=C bond.

The E/Z isomers of but-2-ene are shown in Figure 1. In but-2-ene, the H groups can be on opposite sides of the C=C bond or on the same side of the C=C bond. See later in this topic for details of the Cahn–Ingold–Prelog (CIP) rules, used for assigning E or Z.

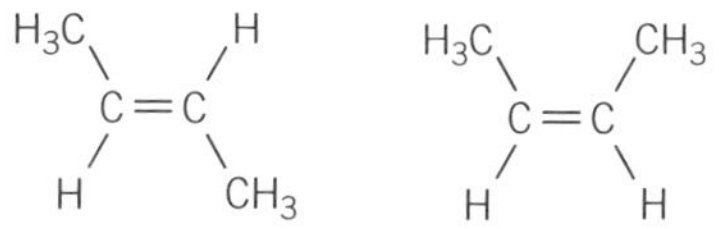

▲ **Figure 1** *The E and Z isomers of but-2-ene with cis and trans labels*

cis–trans isomerism

cis–trans isomerism is a special case of E/Z isomerism in which one of the groups attached to each carbon atom in the C=C bond is the same.

For but-2-ene (Figure 1):

- The *cis* isomer has both hydrogen atoms on the same side of the double bond.
- The *trans* isomer has both hydrogen atoms on opposite sides of the bond.

> **Revision tip**
>
> If an alkene meets the two conditions for E/Z isomerism and there is an H atom attached to each C atom in the C=C bond:
>
> - The *cis* isomer is the Z isomer.
> - The *trans* isomer is the E isomer.

The Cahn–Ingold–Prelog (CIP) priority rules

The CIP rules are used to classify alkenes as E or Z isomers. The atoms attached to each C atom of the C=C bond are given a priority depending on atomic number.

- The atom with the greater atomic number has the higher priority.
- A Z isomer has both higher priority groups on the same side of the C=C bond.
- An E isomer has the higher priority groups on opposite sides of the C=C bond.

Alkenes

Worked example: Assigning E/Z isomers using CIP rules

Figure 2 shows the E and Z isomers of 1-bromo-1-chloropropene.

Figure 2 *The E and Z isomers of 1-bromo-1-chloropropene*

- The left-hand C atom is bonded directly to Br and Cl atoms. Br (at. no. 35) has a greater atomic number than Cl (at. no. 17). Br has the higher priority.
- The right-hand C atom is bonded directly to C and H atoms. C (at. no. 6) has a greater atomic number than H (at. no. 1). C has the higher priority.

In the *E* isomer, the two higher priority groups (Br and C) are arranged on opposite sides of the C=C bond.

In the *Z* isomer the two higher priority groups (Br and C) are on the same side of the C=C bond.

What happens if you have the same priority?

If there are different groups attached to a C atom of the C=C bond, the first atom attached may the same. You will then need to continue along the chains until you find a difference.

In Figure 3, you have to move along the groups on the right-hand carbon until you reach the first difference:

- Cl (at. no. 17) has a greater atomic number and higher priority than O (at. no. 8)
- The two higher priority groups, CH_3 and CH_2CH_2Cl are on the same side of the C=C and the structure is a *Z* isomer.

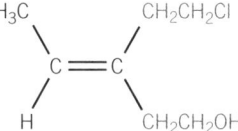

▲ **Figure 3** *Assigning group priorities along a chain*

Summary questions

1 What features does a molecule need for *E/Z* stereoisomerism to occur? *(2 marks)*

2 One of the structural isomers of dichloroethene shows *E/Z* stereoisomerism:
   a Draw and label the stereoisomers as *E/Z* and *cis–trans*. *(2 marks)*
   b Explain why the other structural isomer cannot form *E* and *Z* stereoisomers. *(2 marks)*

3 a Draw structures to show the *E/Z* isomers of 1-bromo-1,2-dichloropropene. Label the isomers *E/Z* and *cis–trans*. *(3 marks)*
   b Explain why the *Z* isomer is *trans*. *(2 marks)*

13.3 Reactions of alkenes

Specification reference: 4.1.3

> **Synoptic link**
>
> You have learnt about addition reactions in Topic 11.5, Introduction to reaction mechanisms.

> **Revision tip**
>
> You need to know the addition reactions of alkenes with:
> - hydrogen in the presence of a nickel catalyst
> - halogens
> - hydrogen halides
> - steam in the presence of an acid catalyst.

> **Synoptic link**
>
> You will learn about the mechanism for this reaction in Topic 13.4, Electrophilic addition in alkenes.

> **Summary questions**
>
> 1. How could you test for unsaturation in a hydrocarbon? *(2 marks)*
>
> 2. Write equations, and state essential conditions for the conversion of ethene into:
>   - **a** an alkane *(2 marks)*
>   - **b** an alcohol *(2 marks)*
>   - **c** a chloroalkane. *(1 mark)*
>
> 3. Butan-2-ol can be prepared from two different structural isomers of C_4H_8.
>   - **a** What are the structural formulae of the isomers? *(2 marks)*
>   - **b** Explain why the purity of the butan-2-ol prepared from each isomer might be different, and suggest the formula of the impurity present. *(2 marks)*

Addition reactions of alkenes

Alkenes undergo many addition reactions. A small molecule adds across the double bond, causing the π-bond to break and new σ-bonds to form.

During addition to alkenes, the double bond is lost: unsaturated → saturated.

Addition reaction with hydrogen

Alkenes react with hydrogen, in the presence of a nickel catalyst, to form alkanes.

For example $\quad C_2H_4 + H_2 \rightarrow C_2H_6$

Addition reaction with halogens

Alkenes react with halogens to form dihaloalkanes.

For example $\quad C_2H_4 + Br_2 \rightarrow BrCH_2CH_2Br$

Test for unsaturation

Unsaturated hydrocarbons decolourise bromine water:

- Orange bromine water is added to an excess of an organic compound and the mixture is shaken.
- In the presence of a double bond, the mixture becomes colourless. The bromine adds across the C=C double bond, forming a dibrominated organic product, which is colourless.
- With a saturated compound, no addition reaction takes place and there is no colour change.

Addition reaction with hydrogen bromide

Alkenes react with hydrogen bromide to form bromoalkanes.

For example $\quad C_2H_4 + HBr \rightarrow CH_3CH_2Br$

Ethene is a symmetrical alkene, with the same groups attached to each carbon atom of the C=C bond. It does not matter which way round the HBr adds across the C=C bond as the product is always the same.

Propene is an unsymmetrical alkene, with different groups attached to each carbon atom of the C=C bond. HBr can add across the C=C bond in two different ways, to produce two products:

$$CH_3CH=CH_2 + HBr \rightarrow CH_3CH_2CH_2Br$$
$$\text{1-bromopropane}$$
$$CH_3CH=CH_2 + HBr \rightarrow CH_3CHBrCH_3$$
$$\text{2-bromopropane}$$

Addition reaction with steam

Alkenes react with steam to form alcohols.

An acid catalyst, such as phosphoric acid, H_3PO_4, must be present.

For example $\quad C_2H_4 + H_2O \rightarrow CH_3CH_2OH$

With an unsymmetrical alkene, two products are possible:

$$CH_3CH=CH_2 + H_2O \rightarrow CH_3CH_2CH_2OH$$
$$\text{propan-1-ol}$$
$$CH_3CH=CH_2 + H_2O \rightarrow CH_3CHOHCH_3$$
$$\text{propan-2-ol}$$

13.4 Electrophilic addition in alkenes

Specification reference: 4.1.3

Electrophilic addition

Alkenes take part in addition reactions to form saturated compounds.

The high electron density of the π-bond in the C=C attracts electrophiles.

- An electrophile is an atom or group of atoms that accepts an electron pair from an electron-rich centre.
- An electrophile is usually electron deficient and may be a positive ion or a molecule containing an atom with a partial positive (δ+) charge.

The mechanism for the reaction of an alkene with an electrophile is called **electrophilic addition**.

> **Key term**
>
> **Electrophile:** A species that can accept a pair of electrons to form a new covalent bond.

Electrophilic addition of hydrogen bromide to an alkene

The equation for the addition of but-2-ene to HBr is shown below:

$$CH_3CH=CHCH_3 + HBr \rightarrow CH_3CH_2CHBrCH_3$$

Electrophilic addition mechanism

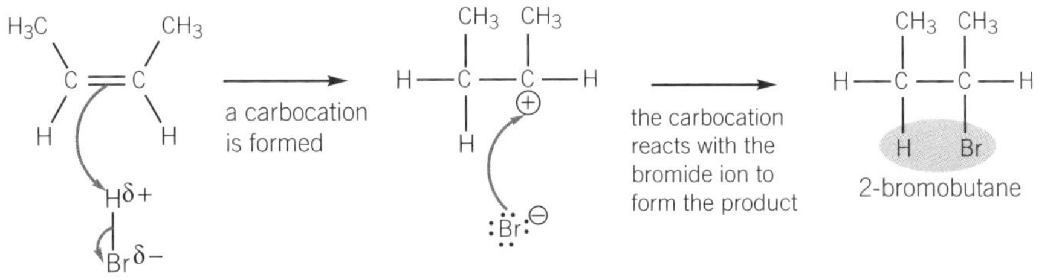

▲ **Figure 1** *The mechanism for the electrophilic addition of but-2-ene with hydrogen bromide*

> **Synoptic link**
>
> You have learnt about curly arrows and reaction mechanisms in Topic 11.5, Introduction to reaction mechanisms.

In the first step:

- H–Br acts as an electrophile, accepting the pair of electrons from the π-bond of C=C.
- The H–Br bond breaks by heterolytic fission.
- A carbocation and a Br⁻ ion are formed.

In the second step:

- The bromide ion is attracted to the carbocation, forming a C–Br bond.
- The 2-bromobutane addition product is formed.

> **Key terms**
>
> **Reaction mechanism:** A series of steps that shows how a reaction takes place.
>
> **Carbocation:** A positively charged species with the positive charge on a carbon atom.

Electrophilic addition of bromine to an alkene

The equation for the addition of propene to bromine is shown below:

$$CH_3CH=CH_2 + Br_2 \rightarrow CH_3CHBrCH_2Br$$

Electrophilic addition mechanism

A bromine molecule is non-polar.

- Initially, the Br_2 molecule approaches the alkene. The high electron density of the π-bond induces a dipole on the Br_2 molecule: $Br^{\delta+}$–$Br^{\delta-}$.
- The Br_2 can now act as an electrophile, with the mechanism (see Figure 2) being similar to the mechanism of HBr with an alkene.

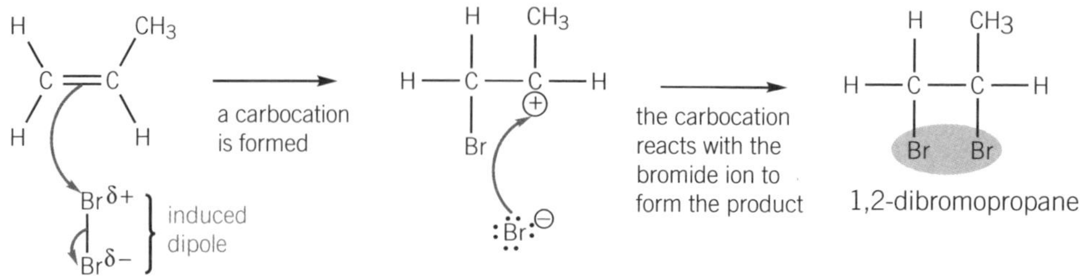

Alkenes

▲ **Figure 2** *The mechanism for the electrophilic addition of but-2-ene with hydrogen bromide*

Addition to unsymmetrical alkenes

In Topic 13.3, you saw that addition of HBr across the double bond of an unsymmetrical alkene (e.g. propene, $CH_3CH=CH_2$) forms two products:

- Addition of HBr to $CH_3CH=CH_2$ forms a mixture of $CH_3CH_2CH_2Br$ and $CH_3CHBrCH_3$.
- $CH_3CHBrCH_3$ is the major product and $CH_3CH_2CH_2Br$ is the minor product.

In the mechanism, the secondary carbocation is more stable than a primary carbocation and is more likely to form. See Figures 3 and 4.

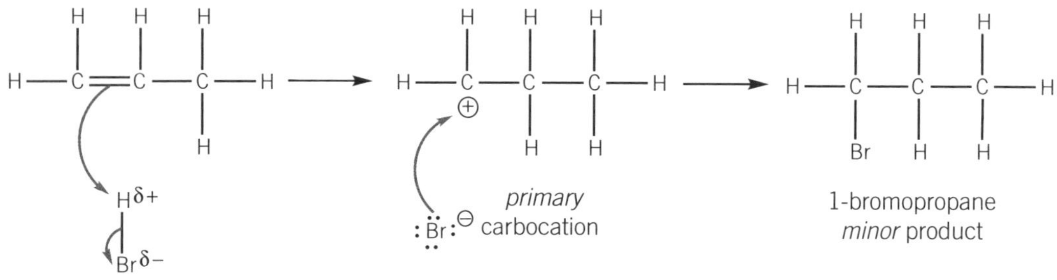

▲ **Figure 3** *Formation of 1-bromopropane goes via a primary carbocation*

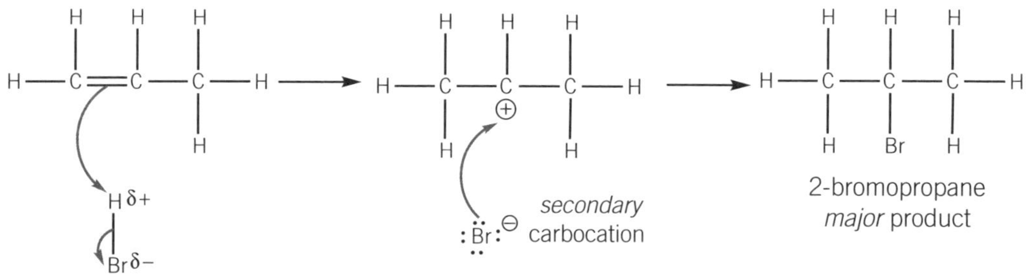

▲ **Figure 4** *Formation of 2-bromopropane goes via a secondary carbocation*

Revision tip
Markownikoff's rule is a useful guide for predicting the major product of an addition reaction of HX with an unsymmetrical alkene.

The stability of carbocations should still be included in any explanation.

Markownikoff's rule

The major product can be predicted using Markownikoff's rule.

- Markownikoff's rule states that when an electrophile H–X reacts with an unsymmetrical alkene, the H atom of H–X attaches to the carbon atom with the greater number of H atoms.

Summary questions

1 What is meant by the terms
   a electrophile   **b** carbocation? *(2 marks)*

2 Draw the mechanism for the reaction of chlorine and ethene. *(3 marks)*

3 Write the structural formulae and names of the two products from the electrophilic addition of HCl to but-1-ene and explain why major and minor products are formed. *(5 marks)*

13.5 Polymerisation in alkenes
Specification reference: 4.1.3

Addition polymerisation

Addition polymers are very large molecules formed by joining together many thousands of alkene molecules (monomers) in 'addition polymerisation'.

- Alkene molecules are unsaturated, containing a C=C bond.
- Polymer molecules are saturated and have no C=C bonds.
- With no C=C bond, polymers are much less reactive than alkenes.

Depending on the monomer used, many different polymers can be formed.

Polymers from monomers

Figures 1 and 2 show equations for the formation of two polymers by addition polymerisation.

- The section drawn in brackets is called a repeat unit.
- The repeat unit is repeated thousands of times in each polymer molecule. The trailing bonds, or 'side links', (bonds through the brackets) represent the covalent bonds joining the repeat units.
- In the equations, there are n monomer molecules and one polymer molecule, with the repeat unit repeated n times.

Monomers from polymers

The monomer used to produce an addition polymer can be identified from the repeat unit of the polymer. First the trailing bonds (side links) are removed. The single C–C is then replaced by a C=C double bond. This is the opposite to the equation for formation of the polymer.

Polymers and the environment

Processing waste polymers

Addition polymers are unreactive and non-biodegradable, persisting for long periods of time, and causing environmental problems.

Waste polymers are processed by various methods.

Recycling

Polymers are collected, sorted, and processed into new polymers.

Use as energy production

Polymers can be burnt. The heat energy produced is used to generate electricity.

Chemical feedstocks

Polymers can be used as an organic feedstock for the production of plastics and other organic products.

Problems with halogenated plastics, e.g. PVC

Combustion of halogenated plastics can form toxic waste products (e.g. HCl gas) which must be removed to prevent discharge into the environment.

Biodegradable polymers

Biodegradable polymers are plant-based and break down naturally to form CO_2, H_2O and biological compounds. When items made from these polymers decompose, there are no toxic waste products to persist in the environment.

Photodegradable polymers

Photodegradable polymers contain bonds that weaken in the presence of light, initiating the breakdown of the polymers.

Synoptic link

For a comparison of the reactivity of alkenes and alkanes see Topic 13.1, The properties of alkenes.

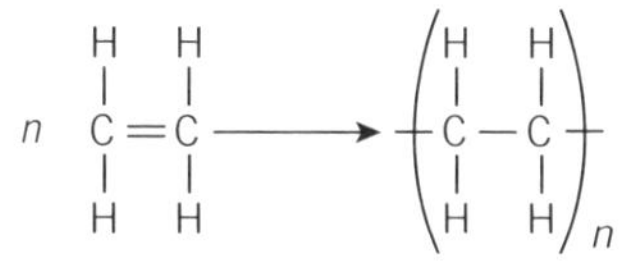

▲ **Figure 1** *The formation of poly(ethene)*

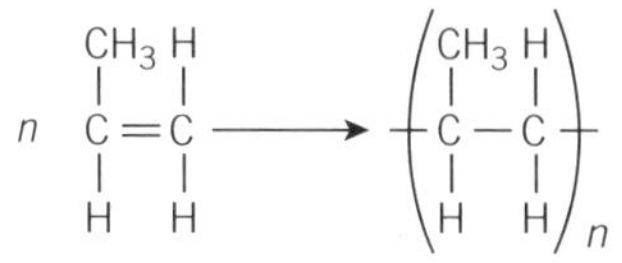

▲ **Figure 2** *The formation of poly(propene)*

Summary questions

1. **a** Draw the repeat unit for the polymer formed from the monomer below. *(1 mark)*

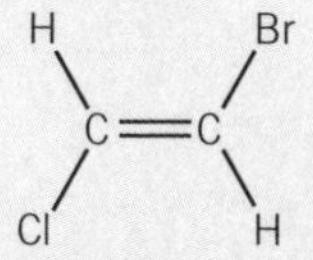

   b Draw the monomer required to make the polymer below. *(1 mark)*

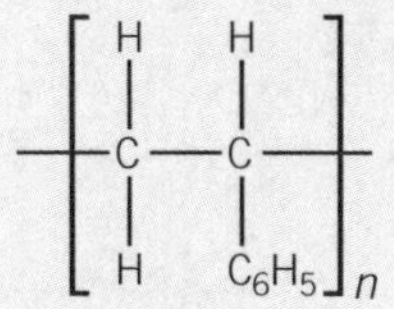

2. Write an equation using displayed formulae, for the addition polymerisation of chloroethene. *(2 marks)*

3. But-1-ene was reacted to make poly(but-1-ene) with an M_r of 15 000.
   a Write an equation for this addition polymerisation. *(2 marks)*
   b Estimate the number of repeat units in poly(but-1-ene). *(2 marks)*

Chapter 13 Practice questions

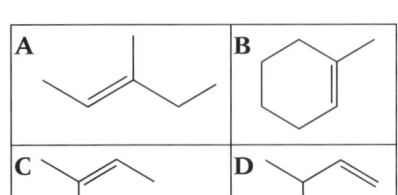

▲ Figure 1

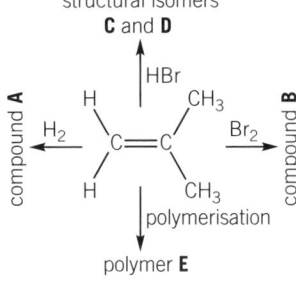

▲ Figure 2

1 What is **not** correct for a π bond in alkenes?
   A a shared pair of electrons
   B stronger than a σ-bond (sigma bond)
   C a sideways overlap of adjacent p orbitals
   D above and below the bonding carbon atoms *(1 mark)*

2 Which compound reacts with HBr to form a compound with $M_r = 164.9$?
   A cyclohexane   B hexane
   C hept-2-ene   D hex-3-ene *(1 mark)*

3 Which alkene in Figure 1 is an *E/Z* stereoisomer? *(1 mark)*

4 What is the mechanism for the reaction between ethene and bromine?
   A electrophilic addition   B electrophilic substitution
   C radical addition   D radical substitution *(1 mark)*

5 This question is about the reactions of propene shown in Figure 2.
   a Show the structures of compounds **A–D**. *(4 marks)*
   b State the catalyst needed for the reaction with H_2.
   c Outline the mechanism for the reaction with HBr to form **C** or **D**. Use your mechanism to explain why the reaction gives a much greater yield of one of the structural isomers. *(5 marks)*
   d i Draw a section of polymer **E**, showing **two** repeat units. *(1 mark)*
      ii Name the type of polymerisation *(1 mark)*
      iii People are concerned about how waste polymers can be disposed of after their useful lifetime. One way is to sort and recycle.
         State **two** other ways that polymers can be processed sustainably instead of being dumped at landfill sites. *(2 marks)*

6 Compound **F** is an *E* stereoisomer.
   a i On analysis, **F** is shown to have the composition by mass of C, 85.71%; H, 14.29% ($M_r = 70$).
      ii Determine the molecular formula of **F**. Show your working.
      iii Identify compound **F** and draw its skeletal formula. Explain your reasoning. *(6 marks)*
   b Reaction of **F** with hydrogen in the presence of a catalyst to form compound **G**.
      i State the catalyst needed for this reaction. *(1 mark)*
      ii Draw the structure of **G**. *(1 mark)*
   c Reaction of **F** with steam in the presence of a catalyst produces a mixture of two structural isomers **H** and **I**.
      i State the catalyst needed for this reaction. *(1 mark)*
      ii Draw skeletal formulae for compounds **H** and **I**. *(2 marks)*
   d Compounds **J** and **K** are *E* and *Z* stereoisomers of 2-bromobut-2-ene. Draw the structures of these stereoisomers and use the Cahn–Ingold–Prelog rules to label each as *E* and *Z*. Explain your reasoning. *(3 marks)*

14.1 Properties of alcohols
Specification reference: 4.2.1

Alcohols
- Alcohols have the general formula $C_nH_{2n+1}OH$.
- Alcohols contain the hydroxyl, –OH, functional group.
- Alcohols with more than two carbon atoms can exist as structural isomers, e.g. propan-1-ol and propan-2-ol (Figure 1). The number indicates the position of the hydroxyl group.

Physical properties of alcohols
Polarity
Alcohols have polar O–H bonds because oxygen is more electronegative than hydrogen. Alcohol molecules are therefore polar and hydrogen bonds exist between the molecules (Figure 2).

Melting and boiling points
The strong hydrogen bonds result in alcohols having relatively high melting points and boiling points, and low volatilities, compared with alkanes of a similar molecular mass.

Solubility
Alcohol molecules can form hydrogen bonds with water molecules. This means that alcohols with a short carbon chain length are soluble in water. Solubility decreases as the carbon chain length increases.

▲ **Figure 1** *Propan-1-ol and propan-2-ol*

▲ **Figure 2** *Hydrogen bonding between alcohol molecules*

> **Synoptic link**
> Hydrogen bonding is covered in detail in Topic 6.4, Hydrogen bonding.

> **Revision tip**
> There are stronger intermolecular forces (hydrogen bonds and London forces) between alcohol molecules than between alkane molecules of a similar molecular mass (London forces only).
>
> This results in alcohols having higher melting and boiling points than alkanes.

Types of alcohol
We can classify alcohols as primary, secondary, or tertiary depending on the number of carbon atoms attached to the carbon atom bonded to the hydroxyl, –OH, group (Figures 3–5).

Primary alcohols
The –OH group is bonded to a carbon which is bonded to one carbon atom.

Methanol, CH_3OH, is the exception with no carbon atoms being attached.

Secondary alcohols
The –OH group is bonded to a carbon which is bonded to two other carbon atoms.

Tertiary alcohols
The –OH group is bonded to a carbon which is bonded to three other carbon atoms.

▲ **Figure 3** *A primary alcohol*

▲ **Figure 4** *A secondary alcohol*

▲ **Figure 5** *A tertiary alcohol*

> **Synoptic link**
> See Topic 13.4, Electrophilic addition in alkenes, for the classification of carbocations as primary and secondary.

Summary questions
1. What is the general formula of an alcohol? *(1 mark)*
2. What type of alcohol is propan-2-ol? *(1 mark)*
3. Why is methanol soluble in water? Your answer should include a diagram. *(3 marks)*

14.2 Reactions of alcohols
Specification reference: 4.2.1

Combustion of alcohols

Alcohols can be burnt to release heat energy. Ethanol is widely used as a fuel.

Complete combustion of ethanol produces carbon dioxide and water.

$$C_2H_5OH(l) + 3O_2(g) \rightarrow 2CO_2(g) + 3H_2O(l)$$

Oxidation of alcohols

Primary and secondary alcohols are oxidised by oxidising agents such as acidified potassium dichromate(VI), $K_2Cr_2O_7/H_2SO_4$.

- Primary alcohols can be oxidised to aldehydes, RCHO, and carboxylic acids, RCOOH.
- Secondary alcohols can be oxidised to ketones, RCOR.
- Aldehydes and ketones both contain the carbonyl group, C=O (Figure 1).

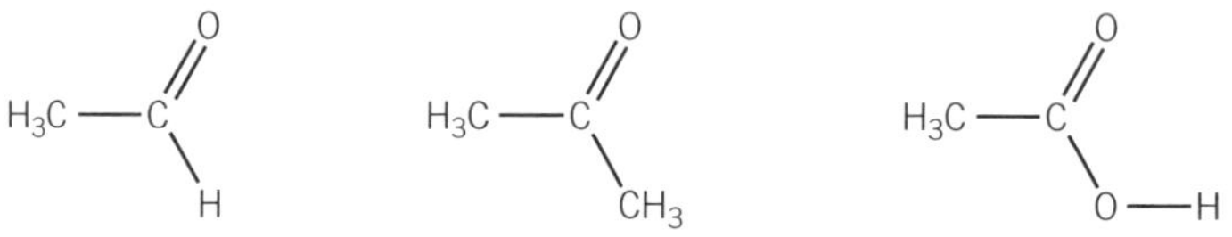

ethanal — an 'aldehyde'
propanone — a 'ketone'
ethanoic acid — a 'carboxylic acid'

▲ **Figure 1** *The aldehyde, ketone, and carboxylic acid functional groups*

Primary alcohols

Oxidation to aldehydes

A primary alcohol is heated with the oxidising agent ($K_2Cr_2O_7/H_2SO_4$) and the organic product is distilled off as it is made. The primary alcohol is oxidised to an aldehyde. Distilling off the aldehyde product prevents further oxidation to a carboxylic acid (see below).

$$CH_3CH_2CH_2CH_2OH + [O] \rightarrow CH_3CH_2CH_2CHO + H_2O$$
butan-1-ol → butanal (an aldehyde)

- Notice that the oxidising agent is written as [O].
- When the primary alcohol is oxidised, chromium in $K_2Cr_2O_7$ is reduced from +6 to +3 and the solution changes colour from orange to green.

Oxidation to carboxylic acids

A primary alcohol is heated under reflux with excess oxidising agent ($K_2Cr_2O_7/H_2SO_4$). Heating under reflux prevents the aldehyde, which initially forms, from escaping and full oxidation to a carboxylic acid can take place.

$$CH_3CH_2CH_2CH_2OH + 2[O] \rightarrow CH_3CH_2CH_2COOH + H_2O$$
butan-1-ol → butanoic acid (a carboxylic acid)

Secondary alcohols

A secondary alcohol is heated under reflux with the oxidising agent ($K_2Cr_2O_7/H_2SO_4$). The secondary alcohol is oxidised to a ketone.

Ketones cannot be oxidised any further.

$$CH_3CH(OH)CH_3 + [O] \rightarrow CH_3COCH_3 + H_2O$$
propan-2-ol → propanone (a ketone)

Revision tip
Notice the suffix in the names:

Aldehydes use the suffix '-al'.

Ketones use the suffix '-one'.

Carboxylic acids use the suffix '-oic acid'.

Synoptic links
You will learn about the reactions of aldehydes, ketones, and carboxylic acids in the second year of the A level course.

See Topic 14.1, Properties of alcohols, for the classification of alcohols as primary, secondary, or tertiary.

Synoptic link
Distillation and heat under reflux are practical techniques that you will need to carry out as part of your practical course. You will meet both in Topic 16.1, Practical techniques in organic chemistry.

Revision tip
Aldehydes and ketones both contain a carbonyl, C=O, functional group.

In aldehydes, the carbonyl group is at the end of the carbon chain: RCHO.

In ketones, the carbonyl group is part of the carbon chain but not at the end: RCOR'.

(R and R' are alkyl groups)

Tertiary alcohols

Tertiary alcohols are **not** oxidised by oxidising agents such as acidified potassium dichromate(VI), $K_2Cr_2O_7/H_2SO_4$. When a tertiary alcohol is heated with $K_2Cr_2O_7/H_2SO_4$, no oxidation reaction takes place and the orange solution stays orange.

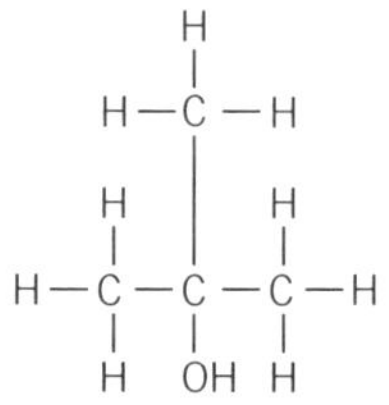

Elimination reactions of alcohols

Dehydrating agents, such as concentrated sulfuric acid, H_2SO_4, or phosphoric acid, H_3PO_4, can be used to remove H_2O from alcohols to form alkenes.

H_2SO_4 or H_3PO_4 acts as an acid catalyst.

For example: $CH_3CH_2CH_2OH \rightarrow CH_3CH=CH_2 + H_2O$

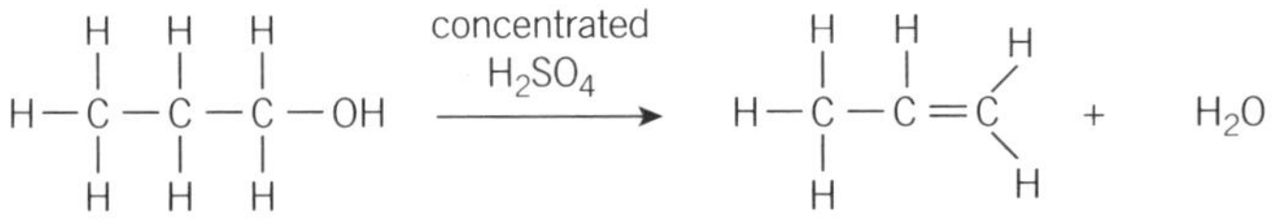

▲ **Figure 3** *Elimination of water from propan-1-ol*

Removal of water from an alcohol is an example of an elimination reaction.

Substitution reactions of alcohols

Alcohols react with halide ions in acidic conditions, e.g. $NaBr/H_2SO_4$, to produce haloalkanes and water. NaBr and H_2SO_4 react together to make hydrogen bromide, HBr, which then reacts with the alcohol.

$CH_3CHOHCH_3 + HBr \rightarrow CH_3CHBrCH_3 + H_2O$

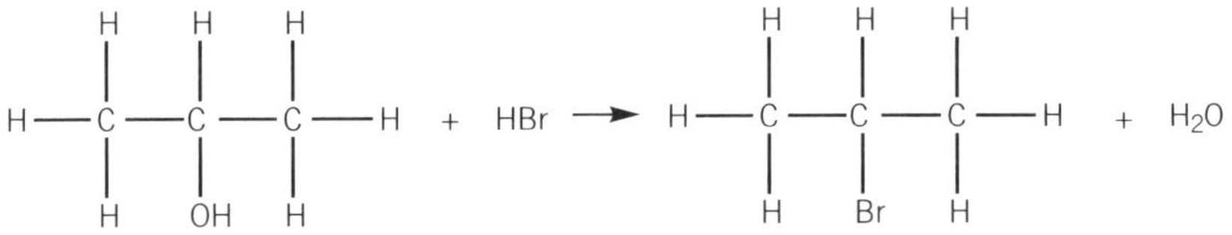

propan-2-ol → 2-bromopropane

▲ **Figure 4** *Substitution of the –OH group in propan-2-ol*

This is an example of a substitution reaction.

▲ **Figure 2** *Methylpropan-2-ol, a tertiary alcohol*

Synoptic link
You have learnt about elimination reactions in Topic 11.5, Introduction to reaction mechanisms.

Synoptic link
You have learnt about substitution reactions in Topic 11.5, Introduction to reaction mechanisms.

Key terms

Elimination reaction: A reaction in which a molecule loses a small molecule, such as water.

Substitution reaction: A reaction in which an atom or group of atoms exchanges with a different atom or group of atoms.

Summary questions

1. Write an equation for the complete combustion of butan-1-ol. *(1 mark)*

2. **a** Write equations, and state essential reagents and conditions, for the conversion of ethanol into:
   i a chloroalkane *(2 marks)*
   ii an aldehyde *(2 marks)*
   iii a carboxylic acid. *(2 marks)*
   b State the types of reaction for the conversions in part **a**. *(2 marks)*

3. Butan-2-ol is heated under reflux with concentrated sulfuric acid. A mixture of three organic compounds forms.
   a Name the type of reaction. *(1 mark)*
   b Write an equation for the reaction, using molecular formulae. *(1 mark)*
   c Name the three organic compounds and draw their skeletal formulae. *(4 marks)*

Chapter 14 Practice questions

1. Which compound is a secondary alcohol?
   - **A** butan-1-ol
   - **B** butan-2-ol
   - **C** 2-methylpropan-1-ol
   - **D** 2-methylpropan-2-ol (1 mark)

2. An alcohol has the structure below.

   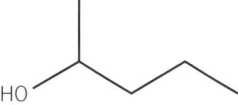

   What is the systematic name for the alcohol?
   - **A** pentan-2-ol
   - **B** pentan-3-ol
   - **C** 1-methylbutan-2-ol
   - **D** 2-methylbutan-2-ol (1 mark)

3. $C_4H_{10}O$ has four structural isomers that are alcohols.
   Which alcohol forms three alkenes on elimination of H_2O?
   - **A** $(CH_3)_2CHCH_2OH$
   - **B** $(CH_3)_3COH$
   - **C** $CH_3CH_2CH_2CH_2OH$
   - **D** $CH_3CH_2CHOHCH_3$ (1 mark)

4. **a** Describe the oxidation reactions of butan-1-ol, using a suitable oxidising agent.

   Your answer should
   - show how different reaction conditions can be used to control the organic product that forms
   - name the functional groups of the organic products formed
   - include reagents, observations and equations.

   In your equations, use structural formulae and use [O] to represent the oxidising agent. (8 marks)

5. Four reactions of alcohol **A** are shown.

   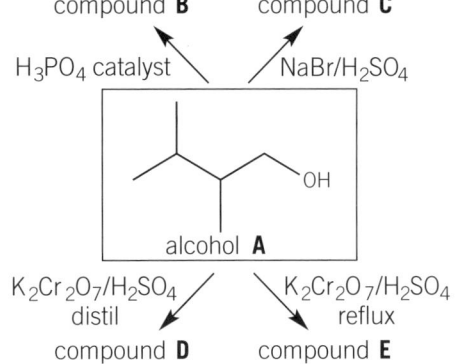

   a Write the systematic name of alcohol **A**. (1 mark)
   b What is the molecular formula of alcohol **A**? (1 mark)
   c Draw skeletal formulae for **B**, **C**, **D**, and **E**. (4 marks)
   d Name the types of reaction for the formation of **B**, **C**, **D**, and **E**. (3 marks)
   e What other product is formed in the formation of compound **B**? (1 mark)
   f Compound **F** is a structural isomer of alcohol **A**.

   Compound **F** reacts in a similar way to alcohol **A** with $K_2Cr_2O_7/H_2SO_4$ and $NaBr/H_2SO_4$ but does **not** react with H_3PO_4.
   Draw the skeletal formula of **F**. (1 mark)

6. A student intends to prepare 10.0 g of 1-bromo-2,2-dimethylbutane starting from 2,2-dimethylbutan-1-ol.

   The student follows an experimental method for this preparation which has a percentage yield of 40.0%.

   a State reagents needed for this reaction. (1 mark)
   b Calculate the mass of 2,2-dimethylbutan-1-ol that the student needs to use.
   Give your answer to **three** significant figures. (3 marks)

15.1 The chemistry of the haloalkanes

Specification reference: 4.2.2

Haloalkanes

- Haloalkanes have the general formula $C_nH_{2n+1}X$. (X is a halogen atom.)
- Haloalkanes contain the halo functional group:

    fluoro, –F   chloro, –Cl   bromo, –Cl   iodo, –I

- As with alcohols, haloalkanes with more than two carbon atoms can exist as structural isomers, e.g. 1-bromopropane and 2-bromopropane.

Reactivity of haloalkanes
Polar bonds

A haloalkane has a polar C–X bond because halogen atoms are more electronegative than carbon (Figure 1). The C–X bond polarity decreases down the group as the electronegativity of the halogen decreases from F → I. The polarity of the C–X bond attracts nucleophiles to the $C^{\delta+}$ atom.

- A nucleophile is an atom or group of atoms that donates an electron pair to an electron-deficient centre.
- A nucleophile is usually electron rich and may be a negative ion, with a lone pair, or a molecule containing an atom with a partial negative ($\delta-$) charge.

This means that haloalkanes are more reactive than alkanes.

The strength of the C–X bond

The strength of the C–X bond also influences the reactivity of haloalkanes. The C–X bond enthalpy decreases down the group.

Table 1 shows that C–Cl, C—Br, and C–I bonds are all weaker than the C–H bonds found in alkanes. This also means that chloro-, bromo-, and iodoalkanes are more reactive than alkanes.

The nucleophilic substitution of haloalkanes

In the nucleophilic substitution of a haloalkane, a nucleophile replaces the halogen in the haloalkane, which is lost as a halide ion.

Hydrolysis with aqueous hydroxide ions, OH⁻(aq)

Hydroxide, OH⁻, ions are nucleophiles with a lone pair of electrons on the oxygen atom, which can be donated to form a new covalent bond. Haloalkanes undergo nucleophilic substitution reactions with aqueous alkali, OH⁻(aq), using NaOH(aq) or KOH(aq), to form alcohols.

For example $CH_3CH_2Cl + OH^- \rightarrow CH_3CH_2OH + Cl^-$

1-chloroethane → ethanol

This reaction can also be described as hydrolysis.

Synoptic link

For details of polarity and electronegativity, see Topic 6.2, Electronegativity and polarity.

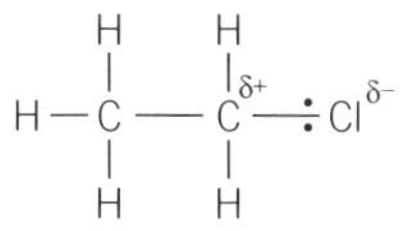

▲ **Figure 1** *Polarity in a carbon–chlorine bond. The bonding pair of electrons is closer to the chlorine atom than the carbon atom*

▼ **Table 1** *Table of bond enthalpy values*

Bond	Bond enthalpy / kJ mol⁻¹
C–F	467
C–H	413
C–Cl	346
C–Br	290
C–I	228

Synoptic link

See Topic 13.4, Electrophilic addition in alkenes, for details of electrophiles.

Electrophiles are the exact opposite of nucleophiles.

Key term

Hydrolysis: A reaction where a substance is split up using water molecules or hydroxide ions.

Haloalkanes

Key term

Nucleophile: A species than donates a pair of electrons to form a new covalent bond. Nucleophiles such as OH^-, CN^-, and NH_3 have a lone pair of electrons which can be donated to another molecule to form a new covalent bond.

Revision tip

When you draw a mechanism, always show curly arrows starting and ending in the correct places.

Curly arrows show the movement of electrons.

- They start from either a lone pair or a bonding pair of electrons.
- They point to the atom to which the electron pair moves.

Synoptic link

See Topic 8.3, Qualitative analysis, for details of the formation of the silver halide precipitates using aqueous silver nitrate.

Nucleophilic substitution mechanism

The mechanism for this hydrolysis is shown below:

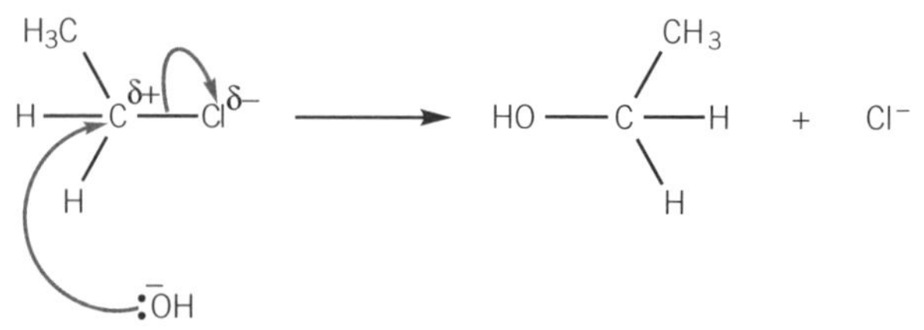

▲ **Figure 2** *The nucleophilic substitution of a haloalkane*

- The OH^- ion acts as a nucleophile, donating an electron pair to the δ+ carbon atom.
- A new bond forms between the oxygen atom of the OH^- ion and the carbon atom.
- The carbon–halogen bond breaks by heterolytic fission.
- The new organic product is an alcohol. A halide ion is also formed.

Hydrolysis with water

Haloalkanes can also be hydrolysed by water in the presence of aqueous silver nitrate to form alcohols. Haloalkanes do not dissolve in water and ethanol is used as a solvent to allow the reactants to mix and to react together.

Water acts as the nucleophile.

For example $\quad CH_3CH_2CH_2X + H_2O \rightarrow CH_3CH_2CH_2OH + H^+ + X^-$

$\qquad$ haloalkane $\qquad \rightarrow \qquad$ alcohol $\qquad$ halide ion

The halide ions react with silver ions to form a silver halide precipitate:

$$Ag^+(aq) + X^-(aq) \rightarrow AgX(s)$$
$$\text{silver halide}$$

- The rates of hydrolysis of different carbon–halogen bonds can be compared by measuring the time for the silver halide precipitate to form.
- The results show that iodoalkanes react faster than bromoalkanes, which react faster than chloroalkanes.
- As bond enthalpies decrease from C–Cl to C–I, the results show that bond enthalpy is a more important factor than bond polarity for explaining the reactivity of haloalkanes.

Summary questions

1. **a** State and explain the trend in the bond polarity of the C–X bond down the halogen group. *(2 marks)*

   b State the trend in the bond enthalpy of the C–X bond down the halogen group. *(1 mark)*

2. **a** Draw the mechanism for the alkaline hydrolysis of bromomethane. *(2 marks)*

   b Name the mechanism and the type of bond fission. *(2 marks)*

3. **a** Outline how you could compare the rates of hydrolysis of different haloalkanes. *(2 marks)*

   b Explain the trend in the rates of hydrolysis. *(2 marks)*

15.2 Organohalogen compounds in the environment

Specification reference: 4.2.2

Uses of organohalogen compounds

CFCs have been used as propellants in aerosols, as refrigerants, and in air-conditioning. CFCs were used because of their low reactivity, high volatility, and because they were non-toxic.

The ozone layer

Benefits of the ozone layer

Ozone helps to filter out harmful ultraviolet (UV) radiation. Overexposure to UV radiation can cause skin cancers, cataracts, and damage to crops.

Breakdown of ozone by radicals

In the upper atmosphere, UV radiation provides the energy to form radicals, which then break down ozone.

Breakdown of ozone by chlorine radicals

CFCs can escape into the Earth's atmosphere where they slowly diffuse into the upper atmosphere. In the presence of UV, C–Cl bonds in CFC molecules are broken by homolytic fission to form radicals.

For example $CCl_3F \rightarrow •CCl_2F + Cl•$

The breakdown of ozone takes place by two propagation steps, catalysed by Cl• radicals.

1st step $Cl• + O_3 \rightarrow ClO• + O_2$
2nd step $ClO• + O \rightarrow Cl• + O_2$
Overall: $O_3 + O \rightarrow 2O_2$

A Cl• radical reacts with ozone in Step 1 and is regenerated in Step 2.

Breakdown of ozone by nitrogen oxide radicals

Nitrogen oxides radicals, e.g. •NO, are made when nitrogen and oxygen react together at high temperatures and pressures. These conditions exist inside aircraft engines and during lightning strikes in thunderstorms.

The •NO radical catalyses the decomposition of ozone in a similar way to Cl•.

1st step $•NO + O_3 \rightarrow •NO_2 + O_2$
2nd step $•NO_2 + O \rightarrow •NO + O_2$
Overall: $O_3 + O \rightarrow 2O_2$

A •NO radical reacts with ozone in Step 1 and is regenerated in Step 2.

> **Revision tip**
> Organohalogens are organic molecules that contain halogen atoms. Chloroalkanes are good solvents. Chlorofluoroalkanes, CFCs, are alkanes that have hydrogen atoms substituted by chlorine and fluorine atoms.

> **Key term**
> **Radical:** A species with an unpaired electron.

> **Synoptic link**
> See Topic 11.5, Introduction to reaction mechanisms, for details of homolytic fission.

> **Synoptic link**
> This process is very similar to the propagation steps in the reaction of Cl• radicals with alkanes. See Topic 12.2, Chemical reactions of the alkanes.

Summary questions

1 Where is the ozone layer found and what does it do? *(1 mark)*

2 How are Cl and NO radicals formed in the upper atmosphere? *(2 marks)*

3 Write equations to show how ozone is broken down by:
   a Cl radicals   b NO radicals. *(4 marks)*

Chapter 15 Practice questions

1. Which carbon–halogen bond has the smallest bond enthalpy?

   A C—F **B** C—Cl **C** C—Br **D** C—I *(1 mark)*

2. The equation for the reaction of 1-bromopropane and hydroxide ions is shown below.

   $$CH_3CH_2CH_2Br + OH^- \rightarrow CH_3CH_2CH_2OH + Br^-$$

   Which statement is **not** correct.

   A The reaction is a substitution

   B Hydroxide ions act as a nucleophile

   C The C—Br bond breaks by homolytic fission

   D The reaction is a hydrolysis *(1 mark)*

3. Which statement is **not** correct for the breakdown of ozone by chlorine radicals?

   A Chlorine radicals act as a heterogeneous catalyst

   B Chlorine radicals are formed from CFCs

   C The breakdown requires UV

   D Ozone is broken down into oxygen *(1 mark)*

4. Which compound would cause the most depletion of the ozone layer?

   A CCl_3F **B** CF_4 **C** $CHClF_2$ **D** CH_2F_2 *(1 mark)*

5. A student investigates the rates of hydrolysis of different haloalkanes. The student heats the haloalkanes with aqueous sodium hydroxide.

   a Write an equation for the reaction of 1-bromobutane with hydroxide ions. *(1 mark)*

   b In this reaction, hydroxide ions behave as a nucleophile.

   What is meant by a *nucleophile*? *(1 mark)*

   c Outline the mechanism for this hydrolysis of 1-bromobutane.

   Show curly arrows and relevant dipoles. *(3 marks)*

   d What type of bond fission has taken place? *(1 mark)*

   e The student repeated the experiment with 1-bromobutane and with 1-iodobutane.

   Predict how the rates of hydrolysis would be different in the three experiments. Explain why there is a difference. *(2 marks)*

   f Haloalkanes react with other nucleophiles.

   Predict the structure of the organic product formed from the reactions below.

   i 2-bromopropane with ethoxide ions, $C_2H_5O^-$ *(1 mark)*

   ii 2-chloro-3-iodobutane with cyanide ions, ^-CN *(1 mark)*

6. Radicals produced from CFCs have been responsible for depletion of the ozone layer.

   a Why is depletion of the ozone layer an environmental concern? *(1 mark)*

   b 1,1,2-Trichloro-1,2,2-trifluoroethane is an example of a CFC, known as Freon 113.

   i Draw the displayed formula for a Freon 113 molecule. *(1 mark)*

   ii How are radicals formed from CFCs? In your answer, include an equation for the production of radicals from Freon 113. *(2 mark)*

   iii Write equations to show how these radicals are able to breakdown ozone and write an overall equation for this process. *(3 marks)*

16.1 Practical techniques in organic chemistry

Specification reference: 4.2.3

Preparation of organic liquids

Organic preparations routinely use Quickfit apparatus and the practical techniques of heating under reflux and distillation.

Heating under reflux

Heating under reflux is vaporisation followed by condensation back to the original container (Figure 1). Using reflux, reactants can be heated safely for a long period of time without the reaction mixture boiling dry.

Distillation

Distillation is used to separate a liquid from a solution or from a mixture of liquids. Distillation is vaporisation followed by condensation to a different container. It can be used to purify a liquid product by removing unreacted reagents and unwanted by-products. Different liquid components can be separated by collection at different boiling points (Figure 2).

Purification of an organic liquid

Removing impurities from organic liquids

Organic liquids can be contaminated with water. If the organic liquid and water are immiscible, the organic liquid forms one layer and the water forms another layer. The denser layer will be at the bottom. The two layers are separated by adding the mixture to a separating funnel and carefully running off each layer though the tap into different containers.

Acid impurities can be removed by adding a solution of sodium carbonate and shaking the mixture formed. Any acid impurities react with sodium carbonate. The organic layer can then be separated from the aqueous sodium carbonate layer using the separating funnel.

Drying the organic product

Even after using a separating funnel, there will still be traces of water contaminating the organic product. The water is removed from the organic product by swirling the mixture with a drying agent. An anhydrous salt is usually used, such as anhydrous magnesium sulfate or anhydrous calcium chloride. The anhydrous salt takes in the water, forming the hydrated salt. The organic liquid is then decanted from the solid hydrated salt into another flask.

Redistillation

If two organic liquids have very similar boiling points, it may be necessary to distil the product a second time to ensure that only the pure liquid is collected.

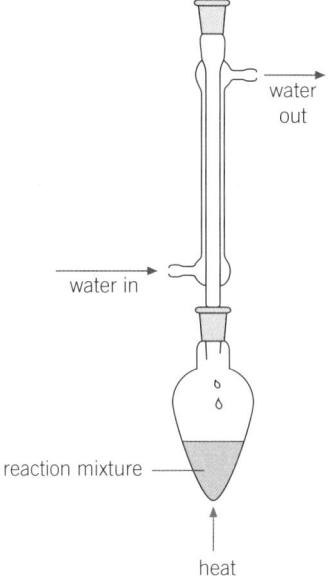

▲ **Figure 1** *Heating under reflux*

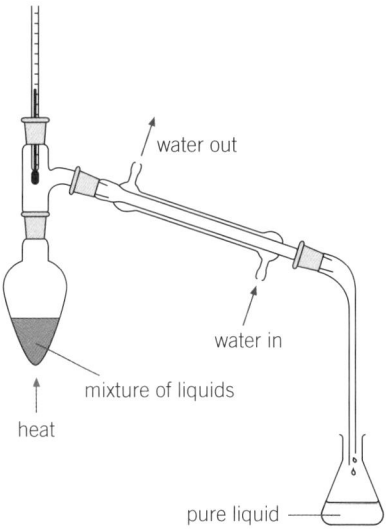

▲ **Figure 2** *Distillation*

Summary questions

1. What is the difference between reflux and distillation? *(2 marks)*

2. Why might a chemist need to redistill a product? *(1 mark)*

3. Describe the stages used to purify an organic liquid contaminated with water to obtain a 'dry' sample. *(3 marks)*

16.2 Synthetic routes
Specification reference: 4.2.3

Development of synthetic routes

Identification of functional groups

Organic chemicals are prepared by converting one functional group into another. When planning a synthesis, it is essential that you can identify functional groups and that you know the key reactions for converting between different functional groups.

> **Synoptic link**
>
> A good knowledge of the functional groups and their reactions is essential when studying organic synthesis. Important functional groups are shown in Topic 11.2, Nomenclature of organic compounds and Topic 14.2, Reactions of alcohols.

> **Worked example: Identifying functional groups**
>
> Two naturally occurring organic compounds are shown below.
>
> The functional groups have been identified and labelled.
>
>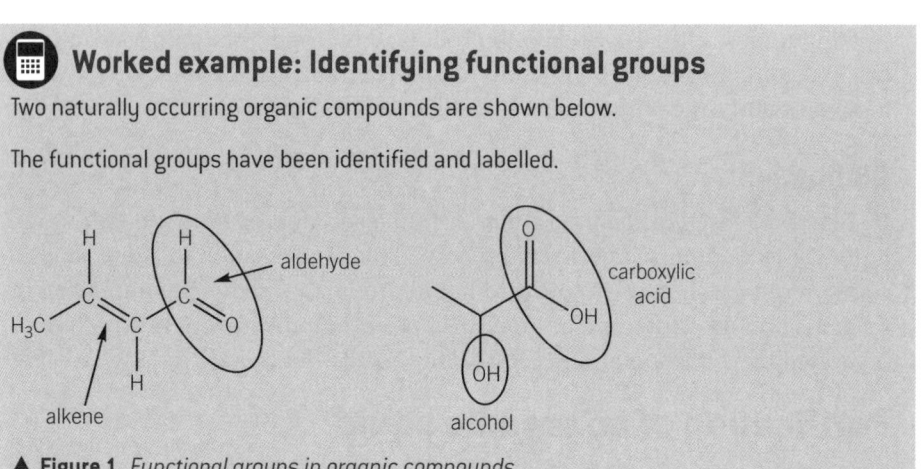
>
> ▲ Figure 1 *Functional groups in organic compounds*

Predicting properties and reactions

You need to learn the reagents and conditions for all the reactions of the functional groups that you have studied: alkanes, alkenes, alcohols, and haloalkanes. The reactions are summarised below.

Alkanes

Alkanes are saturated hydrocarbons with the general formula C_nH_{2n+2}.

▲ Figure 2 *Reactions of alkanes*

Alkenes

Alkenes are unsaturated hydrocarbons with the general formula C_nH_{2n}.

Alcohols

Alcohols have the general formula $C_nH_{2n+1}OH$.

> **Revision tip**
>
> Try writing out the reagents and conditions for the reactions in the course yourself on a blank sheet of paper. If you don't learn this toolkit of reactions, you will struggle to put together a synthetic route.

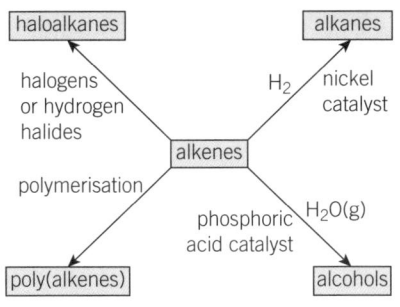

▲ Figure 3 *Reactions of alkenes*

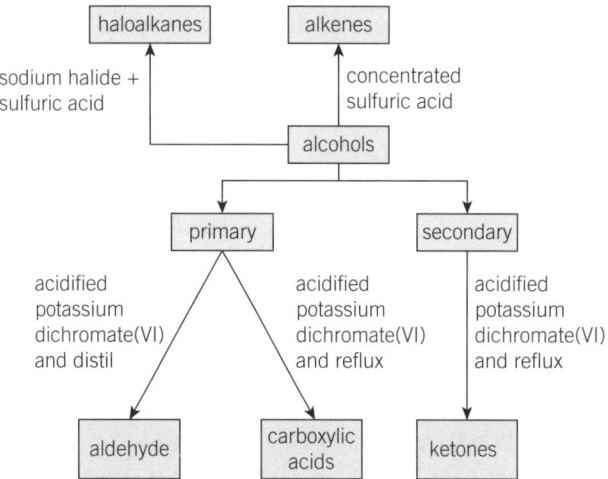

▲ Figure 4 *Reactions of alcohols*

Organic synthesis

Haloalkanes

Haloalkanes have the general formula $C_nH_{2n+1}X$ (X = F, Cl, Br, I).

Constructing two-stage synthetic routes

A plan for a two-stage synthesis is shown below.

- First identify the functional groups in the starting organic material and the final organic product.
- Then work out the intermediate that links the two stages of the synthetic route using reagents and conditions that you have learnt.

▲ **Figure 6** *A two-stage synthesis*

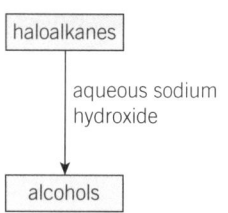

▲ **Figure 5** *Reactions of haloalkanes*

Worked example: Constructing a two-stage synthesis

Construct a two-stage synthesis of butanal starting from 1-bromobutane.

State the reagents and conditions for each stage and write equations for the reactions.

Step 1: Add the formulae of the starting material and the final product to the plan.

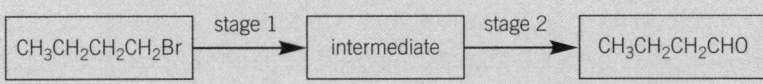

Step 2: Choose a suitable intermediate.

Looking at the toolkit of reactions, butan-1-ol is a suitable intermediate.

Step 3: Finally, write the equation and the reagents and conditions for each stage.

Stage 1

Equation: $CH_3CH_2CH_2CH_2Br + NaOH \rightarrow CH_3CH_2CH_2CH_2OH + NaBr$
Reagent: NaOH(aq)
Conditions: reflux

Stage 2

Equation: $CH_3CH_2CH_2CH_2OH + [O] \rightarrow CH_3CH_2CH_2CHO + H_2O$
Reagent: acidified potassium dichromate(VI)
Conditions: distil

Revision tip

An **intermediate** is a species that is made in the first step and then used up in the second step. Identifying the intermediate is the hardest part of a two-stage synthesis.

Summary questions

1. Name the functional groups in compounds **A** and **B** below. *(2 marks)*

   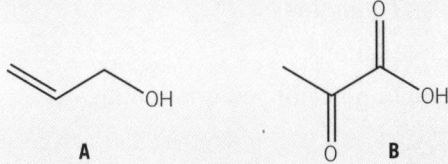

2. How could you prepare 2-bromobutane from the following in one step? For each preparation, write an equation and state the reagents and essential conditions:
   a. an alkene *(1 mark)*
   b. an alcohol. *(2 marks)*

3. Suggest how a sample of propanone can be obtained from propene. Give the equations for the two stages of synthesis and state the reagents and conditions. *(4 marks)*

Chapter 16 Practice questions

1. 1-Bromobutane, $CH_3(CH_2)_3Br$ has a density of $1.27\,g\,cm^{-3}$ and a boiling point range of 101–103 °C

   A student plans to prepare 1-bromobutane by reacting butan-1-ol, $CH_3(CH_2)_3OH$, with sodium bromide and sulfuric acid.

   a   Write an equation for this reaction. *(1 mark)*

   b   The student wants to prepare 5.00 g of 1-bromobutane. The reaction typically produces a 55% yield.

      i   Calculate the mass of butan-1-ol that the student should use to produce 5.00 g of 1-bromobutane.

        Give your answer to **three** significant figures. *(3 marks)*

      ii   The sodium bromide and sulfuric acid are used in excess.

        Calculate the minimum mass of sodium bromide and sulfuric acid that the student needs to use. *(2 marks)*

   c   The reaction is carried out by heating under reflux for 30 minutes.

      i   What is meant by *heating under reflux*? *(1 mark)*

      ii   How would you arrange apparatus for heating under reflux? *(1 mark)*

      iii   Suggest **two** reasons why reflux is used for this reaction. *(2 marks)*

   d   After heating under reflux, the mixture is allowed to cool. An organic layer and an aqueous layer separate out.

      Describe how you would separate the organic layer from the aqueous layer. *(2 marks)*

   e   The organic layer is then dried.

      Explain how you would dry the organic layer, including the name of a suitable drying agent. *(1 mark)*

   f   The 1-bromobutane is contaminated with another organic liquid.

      Explain how you would separate pure 1-bromobutane from the impure mixture of liquids. *(2 marks)*

2. A chemist intends to synthesise two compounds.

   a   The chemist plans to convert compound **A** into compound **B**.

      i   What are the functional groups in **A** and **B**? *(2 marks)*

      ii   Devise a two-stage synthesis for converting compound **A** into **B**.

        Include reagents, conditions and equations.

        Use skeletal formulae in your equations. *(5 marks)*

   b   The chemist plans to convert compound **C** into compound **D**.

      i   What are the functional groups in **C** and **D**? *(2 marks)*

      ii   Devise a two-stage synthesis for converting **C** into **D**.

        Include reagents, conditions, and equations. *(5 marks)*

17.1 Mass spectrometry

Specification reference: 4.2.4

Mass spectrometry in organic chemistry

Mass spectrometry is widely used in the analysis of organic compounds. The key features of a mass spectrum of an organic compound are:

- the molecular ion peak for determination of molecular mass
- fragment ions for determination of parts of the structure.

The mass spectrum of ethanol, CH_3CH_2OH, is shown in Figure 1.

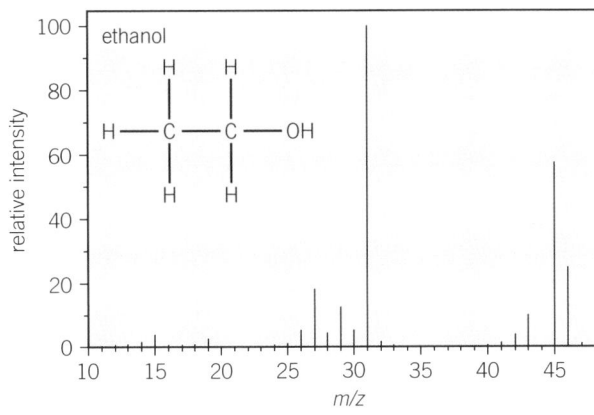

▲ **Figure 1** *The mass spectrum of ethanol*

Molecular ion

The molecular ion peak is the clear peak at the highest m/z value on the right-hand side of the mass spectrum. The molecular mass has the same value as the M peak.

In the mass spectrum of ethanol (Figure 1):

- the molecular ion peak, M, is at $m/z = 46$ for $CH_3CH_2OH^+$
- the molecular mass of ethanol is 46.

Notice the small M+1 peak at $m/z = 47$. The M+1 peak exists because 1.1% of carbon is present as the carbon-13 isotope.

Fragment ions

In the mass spectrometer, some of the molecular ions break down into smaller pieces called **fragment ions**, which are detected as fragment peaks in a mass spectrum.

The mass spectrum of ethanol (Figure 1), contains many fragment ions with m/z values less than $m/z = 46$ for the molecular ion peak.

Some key fragment ions in the mass spectrum of ethanol are listed below:

$m/z = 45$ for $CH_3CH_2O^+$ $m/z = 31$ for CH_2OH^+

Identification of fragment ions can allow chemists to distinguish between different isomers. Even though isomers have the same molecular mass and molecular formula, they have different structures and may fragment in different ways. Table 1 shows some common fragment ions.

Synoptic link

You have already met mass spectrometry in Topic 2.2, Relative mass.

Revision tip

Take care that you do not assign the M+1 peak as the M peak. You will then get the wrong molecular mass!

The M+1 peak is much smaller than the M peak but its size increases as the number of carbon atoms increases.

▼ **Table 1** *Common fragment ions*

m/z	Fragment ion
15	CH_3^+
29	$C_2H_5^+$
43	$C_3H_7^+$ or CH_3CO^+

Revision tip

Fragmentation patterns can be complex. The molecular ion can fragment from either end and fragment ions can themselves also fragment. So use fragment ions with caution.

Spectroscopy

🖩 Worked example: Analysing a mass spectrum

An unknown carboxylic acid has the empirical formula $C_3H_6O_2$. The mass spectrum is shown in Figure 2.

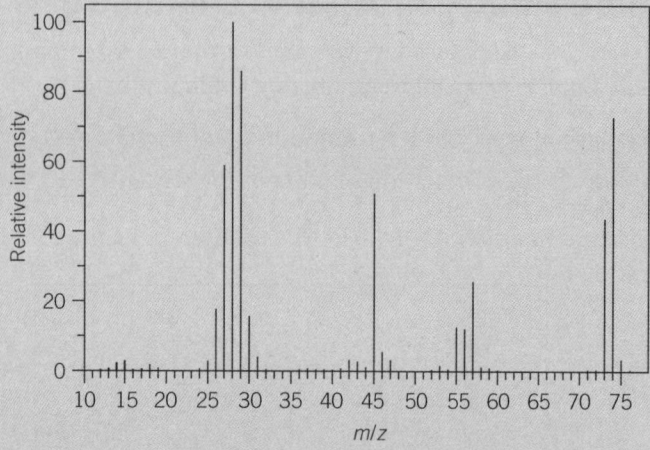

▲ **Figure 2** *The mass spectrum of a carboxylic acid*

What is the molecular formula and possible structure for the carboxylic acid?

Suggest the identity of the fragment ions responsible for the peaks at:

$m/z = 29$; $m/z = 45$; $m/z = 57$; $m/z = 73$

Step 1: Identify the molecular ion peak and a possible structure for the carboxylic acid.

The molecular ion peak is the peak at the highest m/z value on the right at $m/z = 74$.

Empirical formula $C_3H_6O_2$ has a mass of $12 \times 3 + 1 \times 6 + 16 \times 2 = 74$

The molecular mass is 74.

The carboxylic acid must be CH_3CH_2COOH.

Step 2: Identify the fragment ions.

Possible identities of the fragment ions from CH_3CH_2COOH are:

$m/z = 29$, $CH_3CH_2^+$ $m/z = 45$, $COOH^+$

$m/z = 57$, $CH_3CH_2CO^+$ $m/z = 73$, $CH_3CH_2COO^+$

Revision tip
Notice the M+1 peak at $m/z = 75$. If you are not careful, it is easy to mistake this for the M peak.

Summary questions

1. Why is there an M+1 peak in mass spectra? *(1 mark)*

2. The mass spectrum of pentane contains fragment ions at $m/z = 43$ and $m/z = 57$.
   Suggest the formulae of fragment ions causing these peaks. *(2 marks)*

3. The mass spectrum of an alcohol X has the molecular ion peak at $m/z = 60$.
   a. What is the molecular mass and molecular formula for alcohol X? *(2 marks)*
   b. Write structural formulae for the isomers of the alcohol X. *(2 marks)*
   c. The mass spectrum has a large fragment ion at $m/z = 31$.
      Use the fragment ion to suggest, with a reason, which isomer is alcohol X. *(3 marks)*

17.2 Infrared spectroscopy

Specification reference: 4.2.4

Absorption of infrared radiation by atmospheric gases

The covalent bonds in molecules naturally vibrate. In the presence of infrared (IR) radiation, bonds absorb energy and vibrate more. In the atmosphere, IR radiation is absorbed mainly by gases containing C=O, O–H and C–H bonds (e.g. H_2O, CO_2 and CH_4) and re-radiated as heat energy. The concern with global warming is focussed on increased emissions of CO_2 from fossil fuel usage, increasing the absorption of IR and the temperature during re-radiation.

Using IR spectroscopy in organic analysis

Infrared spectroscopy is used to identify the bonds in organic molecules. Different bonds absorb infrared radiation at slightly different wavelengths. We measure the IR radiation using a convenient unit called a wavenumber (cm^{-1}). In an IR spectrum, the wavenumbers of absorption peaks are matched to the bonds in Table 1.

▼ **Table 1** *Characteristic IR absorptions for bonds within functional groups*

Bond	Location	Wavenumber (cm^{-1})
C–C	Alkanes, alkyl chains	750–1100
C–X	Haloalkanes (X = Cl, Br, I)	500–800
C–F	Fluoroalkanes	1000–1350
C–O	Alcohols, esters, carboxylic acids	1000–1300
C=C	Alkenes	1620–1680
C=O	Aldehydes, ketones, carboxylic acids, esters, amides, acyl chlorides and acid anhydrides	1630–1820
aromatic C=C	Arenes	Several peaks in range 1450–1650 (variable)
C≡N	Nitriles	2220–2260
C–H	Alkyl groups, alkenes, arenes	2850–3100
O–H	Carboxylic acids	2500–3300 (broad)
N–H	Amines, amides	3300–3500
O–H	Alcohols, phenols	3200–3600

Revision tip
All organic compounds produce an absorption peak between 2800 and 3100 cm^{-1} for the C–H bond. This is often confused with the O–H peak in alcohols, so you will need to take care.

The fingerprint region

The region of an IR spectrum between 400 cm^{-1} and 1500 cm^{-1} is known as the **fingerprint region**, because it is unique for every compound. An unknown sample can be identified by comparing its infrared spectrum with a database of known IR spectra to find a match in the fingerprint region.

Revision tip
The IR spectrum for an alcohol also shows an absorption at 1000–1300 cm^{-1} for the C–O group. Take care though as this peak is in the fingerprint region.

Interpretation of an IR spectrum

At AS level you should be able to identify the following functional groups from an IR spectrum:

- O–H group in alcohols (from an O–H peak)
- C=O group in aldehydes and ketones (from a C=O peak)
- COOH group in carboxylic acids (from a C=O peak and a broad O–H peak).

Infrared spectrum of an alcohol

Alcohols contain the hydroxyl group, O–H. The infrared spectrum of propan-2-ol contains a peak at 3350 cm^{-1} caused by the O–H bond.

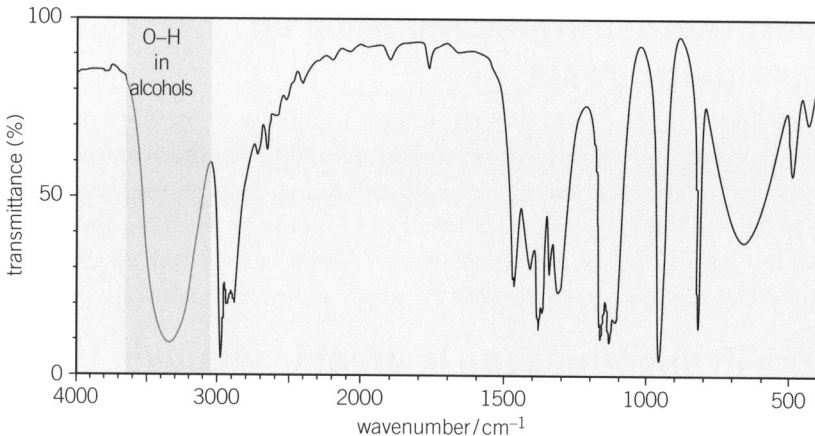

▲ **Figure 1** *The infrared spectrum of an alcohol*

methanol
alcohol

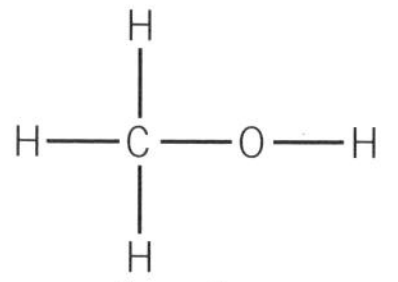

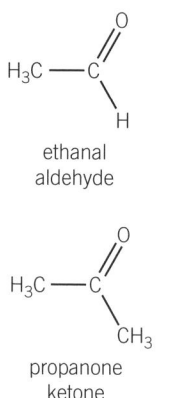

▲ **Figure 2** *Alcohols contain an O–H bond*

Infrared spectrum of an aldehyde or ketone

Aldehydes and ketones both contain the carbonyl group, C=O. The IR spectrum contains an absorption peak for the C=O bond between 1630 and 1820 cm^{-1}.

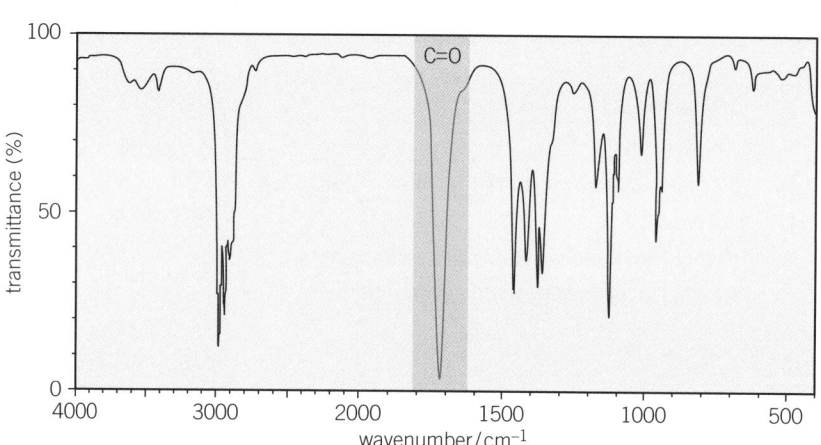

▲ **Figure 3** *The infrared spectrum of an aldehyde or ketone*

ethanal
aldehyde

propanone
ketone

▲ **Figure 4** *Aldehydes and ketones contain a C=O bond*

Infrared spectrum of a carboxylic acid

Carboxylic acids contain the COOH group. The IR spectrum contains an absorption peak between 1630 and 1820 cm^{-1} for the C=O bond and a broad absorption peak between 2500 and 3300 cm^{-1} for the O–H bond.

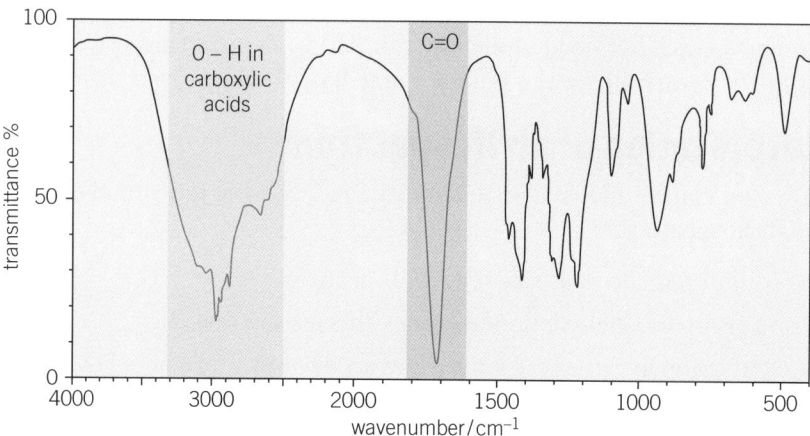

▲ **Figure 5** *The infrared spectrum of a carboxylic acid*

ethanoic acid
carboxylic acid

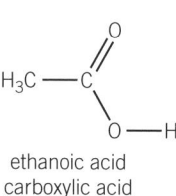

▲ **Figure 6** *Carboxylic acids contain C=O and O–H bonds*

Spectroscopy

Combining techniques

By combining information from a mass spectrum and IR spectrum, the exact identity of the compound can be found:

- Elemental analysis can be used to determine the empirical formula from the percentage composition, by mass, of the elements in a compound.
- Mass spectrometry can be used to identify the molecular ion and therefore the molecular mass of the compound. Used with the empirical formula, the molecular formula can be determined.
- Fragmentation patterns can also be used to deduce the structure of molecules.
- Infrared spectroscopy can be used to identify the bonds and therefore the functional groups in molecules.

Applications of IR spectroscopy

IR spectroscopy has many applications. For example:

- monitoring of gases that cause air pollution (e.g. CO and NO from car emissions)
- in breathalysers to measure ethanol (alcohol) in the breath.

> **Synoptic link**
>
> For details of calculating empirical and molecular formulae, see Topic 2.3, Compounds, formulae, and equations.

Summary questions

1. Would infrared spectroscopy allow a chemist to distinguish between a sample of ethanol and propan-1-ol? Explain your answer. *(1 mark)*

2. Explain how you could distinguish between propan-1-ol, propanal, and propanoic acid from their IR spectra? *(3 marks)*

3. A chemist analyses a sample of an organic compound, which has an empirical formula of C_2H_4O. The infrared spectrum of the compound contains an absorption peak at $1700\,cm^{-1}$ and a broad absorption peak between 2500 and $3300\,cm^{-1}$. The mass spectrum has a molecular ion peak at $m/z = 88$.

   Deduce possible structures for the organic compound. Explain your answer. *(4 marks)*

Chapter 17 Practice questions

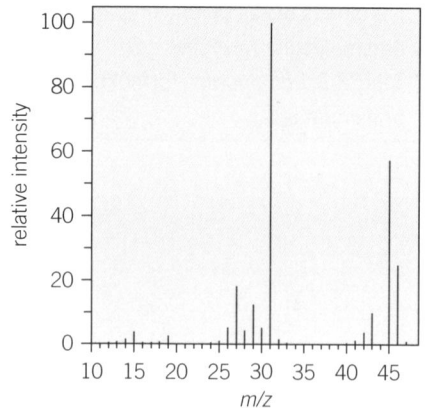

1. Which compound could **not** have produced the molecular ion shown?

   A CH_3CH_2OH

   B $HCOOH$

   C CH_3OCH_3

   D CH_3CHO *(1 mark)*

2. Which alcohol is **not** likely to have a fragment ion at $m/z = 29$ in its mass spectrum?

   A $CH_3CH_2CH(CH_3)OH$

   B $CH_3CH_2CH_2OH$

   C $CH_3CH(OH)CH_2CH_3$

   D $(CH_3)_2CHCH_2OH$ *(1 mark)*

3. Which bond is responsible for the peak labelled **W** in the IR spectrum?

   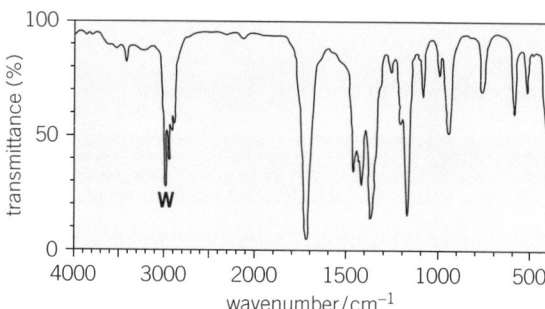

   A C–H

   B O–H in carboxylic acid

   C N–H in amine

   D O–H in alcohol *(1 mark)*

4. Compound **A** is an organic compound containing C, H, and O only. Elemental analysis of compound **A** gave the following percentage composition by mass:

   C, 60.00%; H, 13.33%; O, 26.67%.

   The mass spectrum of compound **A** is shown.

   Compound **A** is heated under reflux with $K_2Cr_2O_7/H_2SO_4$. Compound **B** forms which has the infrared spectrum shown below.

   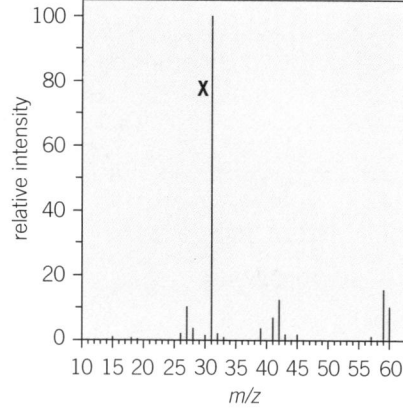

   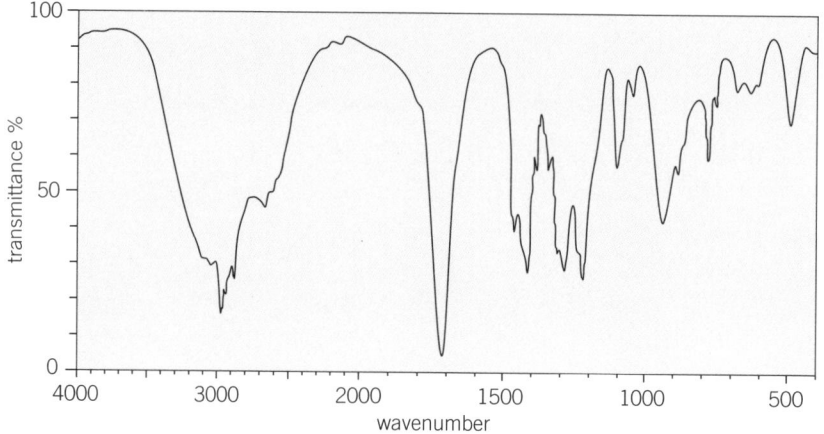

   In your answers, explain fully how you arrive at a structure for compounds **A** and **B** using all the evidence provided.

   - Calculate the empirical and molecular formulae for compound **A**.
   - Identify compounds **A** and **B**. Draw their structures and write an equation for the conversion of **A** into **B**.
   - Write the formula for the particle responsible for peak **X** in the mass spectrum. *(9 marks)*

Answers to summary questions

2.1

1 Protons have a relative mass of 1 and relative charge of +1. Neutrons have a relative mass of 1 and relative charge of 0. [2]

2 They have the same number of electrons / same electron configuration. [1]

3 a 11 protons, 12 neutrons, 11 electrons [1]
   b 26 protons, 30 neutrons, 23 electrons [1]
   c 35 protons, 45 neutrons, 35 electrons [1]
   d 17 protons, 18 neutrons, 18 electrons [1]

2.2

1 a 72.0 [1]
   b 98.0 [1]
   c 46.0 [1]

2 a 94.2 [1]
   b 118.7 [1]
   c 132.1 [1]

3 20.18 [2]

4 Lithium-7
   $(0.0780 \times 6) + (0.922 \times 7) = 6.922$ [2]

2.3

1 a Na_2O [1]
   b Ag_2CO_3 [1]
   c $Ca(OH)_2$ [1]
   d $(NH_4)_2CO_3$ [1]
   e $Cr_2(SO_4)_3$ [1]

2 a $4KClO_3(s) \rightarrow KCl(s) + 3KClO_4(s)$ [1]
   b $2Pb(NO_3)_2(s) \rightarrow 2PbO(s) + 4NO_2(g) + O_2(g)$ [1]
   c $4NH_3(g) + 5O_2(g) \rightarrow 4NO(g) + 6H_2O(l)$ [1]

3 a $CaCO_3(s) + 2HCl(aq) \rightarrow CaCl_2(aq) + CO_2(g) + H_2O(l)$ [1]
   b $2Al(s) + 3Cu(NO_3)_2(aq) \rightarrow 3Cu(s) + 2Al(NO_3)_3(aq)$ [1]
   c $4HCl(aq) + MnO_2(s) \rightarrow MnCl_2(aq) + Cl_2(g) + 2H_2O(l)$ [1]

3.1

1 a $6.02 \times 10^{23} \times 0.300 = 1.806 \times 10^{23}$ [1]
   b $6.02 \times 10^{23} \times 0.750 \times 2 = 9.03 \times 10^{23}$ [1]

2 a $\frac{5.61}{56.1} = 0.100 \text{ mol}$ [1]
   b $0.650 \times 124.0 = 80.6 \text{ g}$ [1]
   c $0.350 \times 78.0 = 27.3 \text{ g}$ [1]

3 a $\frac{12.36}{61.8} = 0.200 \text{ mol}$ [1]
   b $0.150 \times 96.0 = 14.4 \text{ g}$ [1]
   c $n = \frac{4.515 \times 10^{24}}{6.02 \times 10^{23}} = 7.50 \text{ mol}$ [1]
      mass $= 7.50 \times 137.5 = 1031.25 \text{ g}$ [1]

3.2

1 a A crystalline substance that contains water molecules. [1]
   b 6 moles of water of crystallisation per mole of the hydrated salt [1]

2
Element	Fe	O
mass/g	0.279	0.120
$\div A_r$	0.00500	0.00750
$\div$ smallest	1	1.5
$\times 2$ to get whole numbers	2	3

Empirical formula = Fe_2O_3 [2]

3
Compound	$MgSO_4$	H_2O
mass/g	3.16	3.29
$\div M$	0.02625	0.1828
$\div$ smallest	1	7

$MgSO_4 \cdot 7H_2O$ [3]

3.3

1 $\frac{75}{1000} \times 0.0500 = 3.75 \times 10^{-3} \text{ mol}$ [1]

2 $\frac{22.0}{24.0} = 0.917 \text{ mol}$ [1]

3 $V = \frac{nRT}{p} = \frac{2.31 \times 8.314 \times (32 + 273)}{200 \times 10^3} = 0.0293 \text{ m}^3$ [3]

3.4

1 Atom economy = sum of molar masses of desired products / sum of molar masses of all products $\times$ 100
   $\frac{71.0}{153.0} \times 100 = 46.4\%$ [2]

2 Amount of $Na_2O = \frac{1.24}{(23.0 \times 2 + 16.0)} = 0.0200 \text{ mol}$
   Amount of Na $= 0.0200 \times 2 = 0.0400 \text{ mol}$
   Mass of Na $= 0.0400 \times 23.0 = 0.920 \text{ g}$ [2]

3 % yield $= \left(\frac{\text{actual yield}}{\text{theoretical yield}}\right) \times 100$
   Amount of C_2H_4 reacting $= \frac{2.00}{28.0} = 0.0714 \text{ mol}$
   Theoretical mol of $C_2H_5Br = 0.0714 \text{ mol}$
   Actual mol of $C_2H_5Br = \frac{5.80}{108.9} = 0.0533 \text{ mol}$
   % yield $= \left(\frac{0.0533}{0.0714}\right) \times 100 = 74.6\%$ [3]

Answers to summary questions

4 $n(Mg) = \dfrac{0.200}{24.3} = 8.23 \times 10^{-3}$

Ratio Mg : H_2 = 1 : 1

$n(H_2) = 8.23 \times 10^{-3}$ mol

$V = \dfrac{nRT}{p} = \dfrac{(8.23 \times 10^{-3}) \times 8.314 \times 298}{100 \times 10^3}$

$= 2.04 \times 10^{-4}$ m^3

$= 204$ cm^3 [4]

4.1

1 Acids are proton / H^+ donors. [1]

2 a HNO_3 [1]

   b HCl [1]

   c KOH [1]

3 Weak acids are only partially dissociated in water.
   Strong acids are fully dissociated in water. [2]

4 a $KOH + HCl \rightarrow KCl + H_2O$ [2]

   b $2NaOH + H_2SO_4 \rightarrow Na_2SO_4 + 2H_2O$ [2]

   c $CuO + 2HNO_3 \rightarrow Cu(NO_3)_2 + H_2O$ [2]

   d $2CH_3COOH + CaCO_3 \rightarrow (CH_3COO)_2Ca + H_2O + CO_2$ [2]

4.2

1 a $\dfrac{6.00}{(23.0 + 16.0 + 1.0)} = 0.150$ mol dm^{-3} [1]

   b $n(CH_3COOH) = \dfrac{3.20}{60.0} = 0.0533$ mol

   $c(CH_3COOH) = 0.0533 \times \dfrac{1000}{250}$
   $= 0.213$ mol dm^{-3} [2]

   c $\dfrac{21.7}{1000} \times 0.200 = 0.00434$ mol [1]

   d $0.250 \times \dfrac{1000}{2.00} = 125$ cm^3 [1]

2 Amount of $HNO_3 = \dfrac{27.6}{1000} \times 0.150 = 0.00414$ mol
   Amount of NaOH = 0.00414 mol
   Concentration of NaOH = $0.00414 \times \dfrac{1000}{25.0} = 0.1656$
   $= 0.166$ mol dm^{-3} [3]

3 a $Ca(OH)_2(aq) + 2HCl(aq) \rightarrow CaCl_2(aq) + 2H_2O(l)$ [1]

   b $n(HCl) = 2.44 \times 10^{-2} \times \dfrac{25.0}{1000} = 6.10 \times 10^{-4}$ mol
   $n(CaOH)_2 = 0.5 \times 6.10 \times 10^{-4} = 3.05 \times 10^{-4}$ mol
   $c(CaOH)_2 = 3.05 \times 10^{-4} \times \dfrac{1000}{24.20}$
   $= 0.0126$ mol dm^{-3} [3]

4.3

1 Oxidation: loss of electrons and increase in oxidation number.
   Reduction: gain of electrons and decrease in oxidation number. [2]

2 a 0 [1]

   b +1 [1]

   c +6 [1]

3 a Na oxidised from 0 to +1
      O reduced from 0 to −2. [2]

   b Al oxidised from 0 to +3
      H reduced from +1 to 0. [2]

   c Cl oxidised from −1 to 0
      Mn reduced from +4 to +2. [2]

5.1

1 s-orbital: sphere; p-orbital: dumb-bell [1]

2 a $1s^2 2s^2 2p^6 3s^1$ [1]

   b $1s^2 2s^2 2p^6 3s^2 3p^4$ [1]

   c $1s^2 2s^2 2p^6 3s^2 3p^5$ [1]

   d $1s^2 2s^2$ [1]

   e $1s^2 2s^2 2p^6 3s^2 3p^6 3d^6 4s^2$ [1]

   f $1s^2 2s^2 2p^6 3s^2 3p^6 3d^8 4s^2$ [1]

3 a $1s^2 2s^2 2p^6$ [1]

   b $1s^2 2s^2 2p^6 3s^2 3p^6$ [1]

   c $1s^2 2s^2 2p^6$ [1]

   d $1s^2 2s^2 2p^6$ [1]

   e $1s^2 2s^2 2p^6 3s^2 3p^6$ [1]

   f $1s^2 2s^2 2p^6 3s^2 3p^6 3d^{10}$ [1]

5.2

1 The electrostatic attraction between oppositely charged ions. [1]

2 When solid, the ions are in fixed positions in the ionic lattice. When molten, the ions can move as the rigid ionic lattice has broken down. [2]

3 a $Mg^{2+} = 1s^2 2s^2 2p^6$
      $P^{3-} = 1s^2 2s^2 2p^6 3s^2 3p^6$ [2]

   b Mg_3P_2 [1]

4 a $[Ca]^{2+}$ $2[\text{Cl}]^-$ (dot-and-cross diagram) [2]

   b $3[K]^+$ $[N]^{3-}$ (dot-and-cross diagram) [2]

   c $2[Fe]^{3+}$ $3[S]^{2-}$ (dot-and-cross diagram) [2]

5.3

1 a A shared pair of electrons. [1]

   b A shared pair of electrons in which both bonded electrons come from the same atom. [1]

Answers to summary questions

2 a 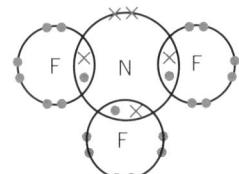 b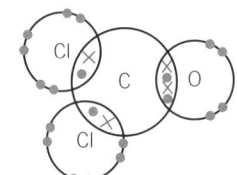

c 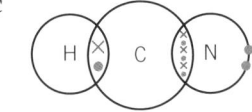   [3]

3 a 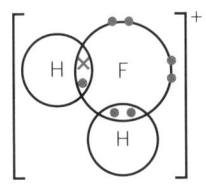 b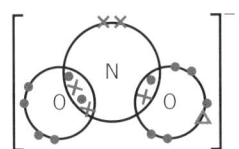

c 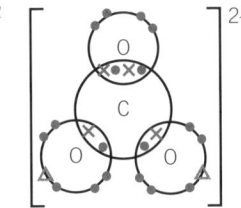   [6]

6.1

1 a The shape is determined by the number of electron pairs around the central atom. Electron pairs repel one another as far apart as possible. [2]

2 a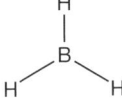

trigonal planar; 120°   [3]

b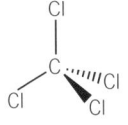

tetrahedral; 109.5°   [3]

c S=C=S

linear; 180°   [3]

d 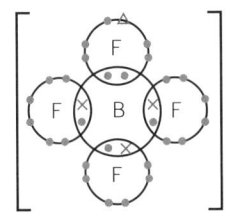...

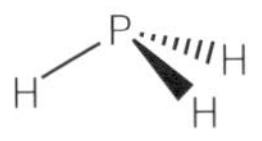

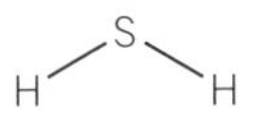

2 a (BH₃) trigonal planar; 120° [3]

b (CCl₄) tetrahedral; 109.5° [3]

c S=C=S linear; 180° [3]

d (PH₃) pyramidal; 107° [3]

e (H₂S) non-linear; 104.5° [3]

3 (BF₄⁻) tetrahedral; 109.5° [3]

6.2

1 Electronegativity is the ability of an atom to attract the bonding electrons in a covalent bond. [1]

2 a $^{\delta+}$P–O$^{\delta-}$ [1]

b $^{\delta+}$Si–Cl$^{\delta-}$ [1]

c $^{\delta-}$N–S$^{\delta+}$ [1]

d no dipole (same electronegativities) [1]

3 a NH_3 is polar: N–H is a polar bond and the molecule is unsymmetrical, so the dipoles do not cancel. [2]

b BF_3 is non-polar: B–F is a polar bond but the molecule is symmetrical and the dipoles cancel. [2]

c CCl_4 is non-polar: C–Cl is a polar bond but the molecule is symmetrical and the dipoles cancel. [2]

d PH_3 is non-polar: P and H have the same electronegativity and there are no polar bonds. [2]

6.3

1 In a polar molecule, the more electronegative atom attracts the less electronegative atom in another polar molecule. [1]

2 Xenon has more electrons and so has stronger induced dipole–dipole interactions/London forces between molecules. [2]

3 The electrons in a molecule are moving causing a temporary dipole. This causes an induced dipole in a neighbouring molecule. Induced dipoles attract, resulting in a force of attraction between the molecules. [2]

4 F_2 has more electrons than O_2 and the London forces between the F_2 molecules will be stronger. F_2 has non-polar molecules and has London forces only. HCl has polar molecules and will also have permanent dipole–dipole interactions. The total intermolecular forces in HCl are stronger. [2]

6.4

1 Ice is less dense than water because the water molecules in ice are held in an open structure by the hydrogen bonds between the molecules. [2]

2 The hydrogen bonds between H_2O molecules increase the strength of the overall intermolecular forces in ice and water. More energy will be needed to break the hydrogen bonds, increasing the melting and boiling point. [2]

3 a 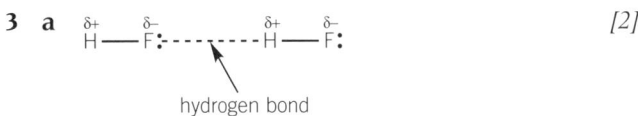 [2]

hydrogen bond

b 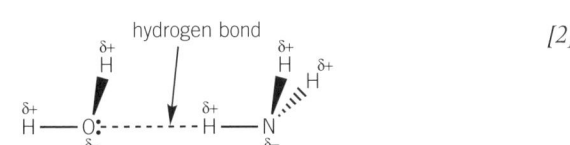 [2]

hydrogen bond

107

Answers to summary questions

7.1

1. The repeating trend in properties across each period in the periodic table. [1]
2. a $1s^22s^22p^1$ p-block [2]
   b $1s^22s^22p^5$ p-block [2]
   c $1s^22s^22p^6$;   p-block [2]
   d $1s^22s^22p^63s^23p^64s^1$ s-block [2]
   e $1s^22s^22p^63s^23p^63d^64s^2$ d-block [2]
3. a $5s^1$ [1]
   b $6s^26p^3$ [1]

7.2

1. First ionisation energy is the energy required to remove one electron from each atom in 1 mole of gaseous atoms to form 1 mole of gaseous 1+ ions. [2]
2. $C^{3+}(g) \rightarrow C^{4+}(g) + e^-$ [1]
3. First ionisation energy increases across a period. The nuclear charge increases and electrons are added to same shell. There is more attraction between the nucleus and the outer electrons, decreasing the atomic radius. [3]
4. In O, the electron is lost from a p orbital that contains paired electrons. In N, all three p orbitals contain one unpaired electron. In O, the paired electrons repel one another and one of the paired electrons is lost more easily than one of the unpaired electrons in N. [2]

7.3

1. a London forces between covalently bonded molecules, simple molecular lattice. [2]
   b Metallic bonding, giant metallic lattice. [2]
2. Sodium to aluminium conduct electricity. They have giant metallic lattices and the delocalised electrons move. Silicon to argon do not conduct electricity. Silicon has a giant covalent lattice and all electrons are fixed in position by covalent bonds. Phosphorus to argon have simple molecular lattices and there are no charged particles. [3]
3. Sulfur forms a simple molecular structure. When sulfur melts, only the weak London forces between molecules are broken. This requires little energy, which is available at low temperatures.

   Silicon forms a giant covalent structure. When silicon melts, the strong covalent bonds between atoms are broken. A large amount of energy is required and this requires high temperatures. [4]
4. Phosphorus has P_4 molecules with more electrons than chlorine with Cl_2 molecules. The London forces between P_4 molecules are stronger than Cl_2 molecules and so the melting point is higher. [2]

8.1

1. Ionisation energy decreases down the group. The atomic radius increases. There is more shielding by inner electrons. There is less attraction between the nucleus and the outer electrons and so the outer electrons are lost more easily. [3]
2. $Sr^+(g) \rightarrow Sr^{2+}(g) + e^-$ [1]
3. $Ba(s) + 2H_2O(l) \rightarrow Ba(OH)_2(aq) + H_2(g)$
   Ba is oxidised from 0 to +2.
   H is reduced from +1 to 0. [3]

8.2

1. Br atom: $1s^22s^22p^63s^23p^63d^{10}4s^24p^5$;
   Br⁻ ion: $1s^22s^22p^63s^23p^63d^{10}4s^24p^6$ [2]
2. Reactivity decreases down the group. Down the group the electron is gained by being placed into a shell which is further from the nucleus. The shielding from inner shells increases. The attraction between the nucleus and the electron decreases and so the electron is gained less easily. [3]
3. a $Cl_2(aq) + 2I^-(aq) \rightarrow 2Cl^-(aq) + I_2(aq)$ [2]
   b Violet [1]
   c Iodine is oxidised from −1 in I⁻ to 0 in I_2.
   Chlorine is reduced from 0 in Cl_2 to −1 in Cl⁻. [2]

8.3

1. $Ba^{2+}(aq) + SO_4^{2-}(aq) \rightarrow BaSO_4(s)$ [1]
2. Add silver nitrate solution. A cream precipitate would form, which dissolves in concentrated $NH_3(aq)$ but not in dilute $NH_3(aq)$. [2]
3. Add sodium hydroxide solution to the sample and warm the mixture.
   If the sample contains ammonium ions, alkaline ammonia gas will be produced, which turns damp red litmus blue. [2]

9.1

1. Any solutions have a concentration of $1.00\,mol\,dm^{-3}$.
   Any gases must have a pressure of 100 kPa.
   A temperature of 298 K/25 °C. [1]
2. Zero: nitrogen is an element and its standard state is $N_2(g)$. [1]
3. $C_2H_5OH(l) + 3O_2(g) \rightarrow 2CO_2(g) + 3H_2O(l)$ [2]
4. $C(s) + 2H_2(g) \rightarrow CH_4(g)$ [2]

Answers to summary questions

9.2

1. Heat loss. Add insulation and use a lid. [2]
2. a $50.0 \times 4.18 \times (57.5 - 21.0) = 7628.5\,J$ or $7.6285\,kJ$ [1]

   b Amount of $CuSO_4 = 1.00 \times \dfrac{50.0}{1000} = 0.0500\,mol$

   Enthalpy change $= \dfrac{7.6285}{0.0500} = -153\,kJ\,mol^{-1}$ [3]

3. $q = 150.0 \times 4.18 \times 64 = 40\,128\,J$ or $40.128\,kJ$

   Amount of $C_3H_7OH = \dfrac{1.50}{60.0} = 0.0250\,mol$

   $\Delta_c H = \dfrac{40.128}{0.0250} = -1605.12 = -1610\,kJ\,mol^{-1}$ (3 s.f.) [4]

9.3

1. The mean amount of energy required to break one mole of a specified type of covalent bond in gaseous molecules. [2]
2. The average bond enthalpy for a bond is an average value obtained from many molecules. The actual bond enthalpy of a particular bond may be slightly different from the average. [1]
3. **Energy required to break the existing bonds:**

   $4 \times C-H = 4 \times 413 = 1652\,kJ\,mol^{-1}$

   $3 \times O=O = 3 \times 498 = 1494\,kJ\,mol^{-1}$

   $1 \times C=C = 1 \times 612 = 612\,kJ\,mol^{-1}$

   Total $= 3758\,kJ\,mol^{-1}$

   Energy released to form new bonds:

   $4 \times C=O = 4 \times 805 = 3220\,kJ\,mol^{-1}$

   $4 \times O-H = 4 \times 464 = 1856\,kJ\,mol^{-1}$

   Total $= 5076\,kJ\,mol^{-1}$

   $\Delta H = 3758 - 5076 = -1318\,kJ\,mol^{-1}$ [3]

9.4

1. $[-110] - [-75 -242] = +207\,kJ\,mol^{-1}$ [2]
2. $[(5 \times -393.5) + (6 \times -285.8)] - [-3509.1] = -173.2\,kJ\,mol^{-1}$ [3]
3. $[(2 \times -602) + (4 \times -33)] - [(2 \times -791)] = +246\,kJ\,mol^{-1}$ [3]

10.1

1. Activation energy is the minimum energy for a reaction to take place. [1]
2. Increasing pressure increases the rate. The particles have a greater concentration and collide more frequently. [2]
3. a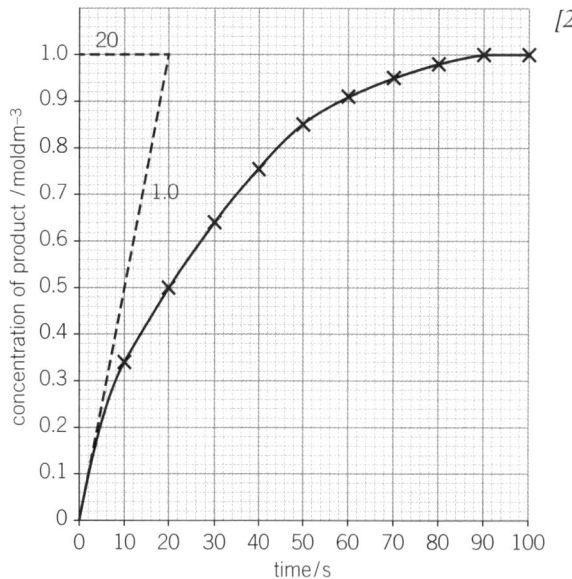

   b Initial rate $= \dfrac{1.0}{20} = 0.050\,mol\,dm^{-3}\,s^{-1}$ [2]

10.2

1. Catalysts provide an alternative reaction pathway with a lower activation energy. More particles will have energy greater than or equal to the lower activation energy, increasing the rate of reaction. [2]
2. Economic – lower temperatures and pressures can be used which will be less expensive to generate.

   Environmental – lower temperatures reduce the energy demand, less fossil fuels are burnt, and less carbon dioxide gas is produced. [2]
3. a A homogeneous catalyst is in the same state as the reactants. In the breakdown of ozone, the reactants $O_3(g)$ and $O(g)$ are gases and the catalyst $Cl(g)$ is a gas. [2]

   b A heterogeneous catalyst is in a different state from the reactants. In the preparation of ammonia, the reactants $N_2(g)$ and $H_2(g)$ are gases and the catalyst $Fe(s)$ is a solid. [2]

10.3

1. The total number of molecules. [1]
2. Adding a catalyst does not change the distribution curve but the activation energy is now at a lower energy. More molecules exceed the new lower activation energy, allowing more molecules to react on collision. The rate of reaction increases. [3]
3. The average energy of the particles decreases. The peak of the distribution curve moves to the left and the distribution curve becomes higher. Fewer molecules exceed the activation energy and the rate of reaction decreases. [3]

Answers to summary questions

10.4

1 The reaction is at equilibrium. [1]

2 The rate of reaction increases because more molecules now exceed the activation energy. The position of equilibrium shifts in the endothermic direction, to the right-hand side, to minimise the temperature increase. [2]

3 a High pressure because there are fewer gaseous molecules on the right. Low temperature because the forward reaction is exothermic. [2]

   b A high pressure requires a large quantity of energy, increasing the cost of the process. High pressures also introduce risks to safety. A low temperature may produce a reaction rate so slow that equilibrium may not be reached. [2]

10.5

1 $K_c = \dfrac{[SO_3(g)]^2}{[SO_2(g)]^2\,[O_2(g)]}$ [1]

2 $K_c = \dfrac{[HI(g)]^2}{[H_2(g)]\,[I_2(g)]} = \dfrac{0.321^2}{0.0520 \times 0.125} = 15.9$ [3]

3 $K_c = \dfrac{[CH_3OH(g)]}{[CO(g)][H_2(g)]^2}$

   $[CH_3OH(g)] = K_c \times [CO(g)] \times [H_2(g)]^2$
   $= 14.6 \times 0.310 \times 0.240^2$
   $= 0.261\,mol\,dm^{-3}$ [3]

11.1

1 a A saturated hydrocarbon contains single carbon–carbon bonds only. [1]

   b An aromatic hydrocarbon contains a benzene ring. [1]

   c A homologous series is a series of organic compounds that has the same functional group. Each successive member of a homologous series differs from the previous member by the addition of CH_2. [2]

2 a C_7H_{16} [1]
   b $C_{25}H_{50}$ [1]
   c $C_6H_{13}Cl$ [1]

3 a C_nH_{2n-2} [1]
   b $C_{12}H_{22}$ [1]

11.2

1 a An alkyl group has the general formula C_nH_{2n+1} with one fewer H atom than the parent alkane. [1]

      i C_9H_{20} [1]
      ii C_7H_{15} [1]

2 a propan-1-ol [1]
   b 2-bromobutane [1]
   c pent-2-ene [1]

   d 2,3-dimethylpentane [1]
   e 1-bromo-2-methylhexane [1]

3 a [1]

   b [1]

   c 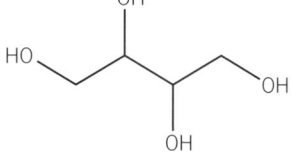 [1]

11.3

1 Molecular: C_4H_8; empirical: CH_2;
   structural: $CH_3CH_2CHCH_2$
   displayed and skeletal:
   [5]

2 Skeletal:

   structural: $HOCH_2CHOHCHOHCH_2OH$
   molecular: $C_4H_{10}O_4$; empirical: $C_2H_5O_2$; [4]

3 Molecular formula: $C_{10}H_{16}$; empirical formula: C_5H_8 [2]

11.4

1 Structural isomers are compounds with the same molecular formula but different structural formulae. [1]

2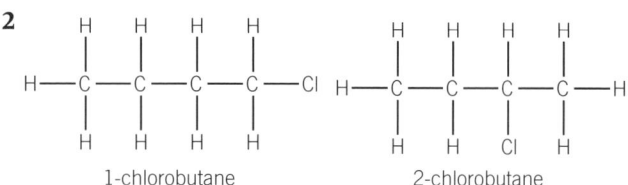

1-chlorobutane   2-chlorobutane

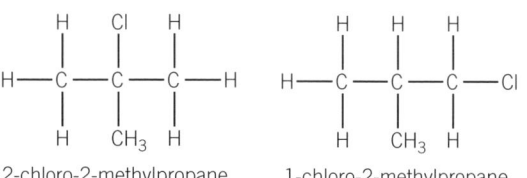

2-chloro-2-methylpropane   1-chloro-2-methylpropane

[8]

Answers to summary questions

3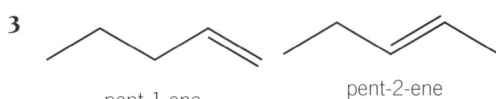

pent-1-ene   pent-2-ene

2-methylbut-1-ene   2-methylbut-2-ene   3-methylbut-1-ene

[5]

11.5

1. A species with an unpaired electron. [1]
2. The movement of a pair of electrons. [1]
3. An addition reaction. [1]

12.1

1. Sigma bonds are formed from the overlap of orbitals directly between the bonding atoms. [1]
2. Each carbon atom is surrounded by four bonded pairs of electrons. The four bonded pairs repel each other equally, resulting in a tetrahedral shape with bond angles of 109.5°. [4]
3. The boiling point increases as the number of carbon atoms increases. There are more electrons and there are also more points of contact between molecules, increases the strength of the London forces and the energy needed to break the intermolecular forces. [3]

12.2

1. a UV radiation [1]
   b Radical substitution [1]
2. Complete: $C_4H_{10} + 6½O_2 \rightarrow 4CO_2 + 5H_2O$

   Incomplete: $C_4H_{10} + 4½O_2 \rightarrow 4CO + 5H_2O$ [2]
3. a i $Cl_2 \rightarrow 2Cl\bullet$ [1]

   ii $Cl\bullet + C_2H_6 \rightarrow C_2H_5\bullet + HCl$

   $C_2H_5\bullet + Cl_2 \rightarrow C_2H_5Cl + Cl\bullet$ [2]

   iii $2Cl\bullet \rightarrow Cl_2$

   $Cl\bullet + C_2H_5\bullet \rightarrow C_2H_5Cl$

   $2C_2H_5\bullet \rightarrow C_4H_{10}$ [3]

   b CH_3CHCl_2, $ClCH_2CH_2Cl$, CH_3CCl_3, $ClCH_2CHCl_2$, $ClCH_2CCl_3$, $Cl_2CHCHCl_2$, Cl_2CHCCl_3, Cl_3CCCl_3 [3]

13.1

1. A π-bond is the sideways overlap of adjacent p orbitals above and below the bonding C atoms. [1]
2. Each carbon is surrounded by three regions of electron density. The three regions repel each other as far apart as possible to gives a trigonal planar shape around each carbon atom with a bond angle of 120°. [4]
3. The high electron density of the π-bond above and below the plane of the C=C bond attracts electrophiles. The π-bond has a smaller bond enthalpy than the σ-bond and is broken more easily. Alkanes have only σ-bonds. [3]

13.2

1. There needs to be a C=C double bond and two different groups attached to each carbon atom of the C=C bond. [2]

2. a [Z cis and E trans structures of C=C with Cl/H groups] [2]

   b The other structural isomer is $CH_2=CCl_2$. One of the carbon atoms of the C=C bond is bonded to two H atoms and the other to two Cl atoms. For E/Z isomerism, both carbons of the C=C bond must be bonded to different groups. [2]

3. a [Z trans and E cis structures with H_3C, Cl, Br groups] [3]

   b In the Z isomer, Cl and Br are higher priority groups on the same side of the C=C bond.

   The Z isomer is *trans* because the common Cl groups are on opposite sides of the C=C bond. [2]

13.3

1. Orange bromine water is added to the organic compound and the mixture shaken. Unsaturated compounds decolourise the bromine water. [2]
2. a $C_2H_4 + H_2 \rightarrow C_2H_6$ Conditions: Ni catalyst [2]

   b $C_2H_4 + H_2O \rightarrow C_2H_5OH$ Conditions: Steam and an acid catalyst. [2]

   c $C_2H_4 + HCl \rightarrow C_2H_5Cl$ [1]

Answers to summary questions

3 a CH$_3$CH=CHCH$_3$ and CH$_3$CH$_2$CH=CH$_2$ [2]

b H$_2$O adds across the C=C in CH$_3$CH=CHCH$_3$ to form one product, butan-2-ol, CH$_3$CH$_2$CHOHCH$_3$.

With CH$_3$CH$_2$CH=CH$_2$, a mixture of butan-2-ol, CH$_3$CH$_2$CHOHCH$_3$, and the impurity butan-1-ol, CH$_3$CH$_2$CH$_2$CH$_2$OH, is formed. [2]

13.4

1 a A species that can accept a pair of electrons to form a new covalent bond. [1]

b A positively charged species with the positive charge on a carbon atom. [1]

2 [mechanism diagram showing addition of Cl–Cl across C=C via carbocation intermediate] [3]

3 2-chlorobutane, CH$_3$CH$_2$CHClCH$_3$, and 1-chlorobutane, CH$_3$CH$_2$CH$_2$CH$_2$Cl.

In the reaction mechanism, the intermediate in the formation of 2-chlorobutane is the secondary carbocation, CH$_3$CH$_2$CH$^+$CH$_3$. For formation of 1-chlorobutane, the intermediate is a primary carbocation, CH$_3$CH$_2$CH$_2$CH$_2^+$. Secondary carbocations are more stable than primary carbocations and so CH$_3$CH$_2$CH$^+$CH$_3$ is more likely to form. Therefore, CH$_3$CH$_2$CHClCH$_3$ will be the major product and CH$_3$CH$_2$CH$_2$CH$_2$Cl will be the minor product. [5]

13.5

1 a [poly(chlorobromoethene-type) repeat unit with H, Br, Cl, H] [1]

b [structure of styrene: CH$_2$=CHC$_6$H$_5$] [1]

2 [chloroethene monomer → poly(chloroethene) repeat unit] [2]

3 a [but-1-ene monomer → poly(but-1-ene) repeat unit] [2]

b M_r for C$_4$H$_8$ = 56

Repeat units = $\dfrac{15\,000}{56}$ = 268 [2]

14.1

1 C$_n$H$_{2n+1}$OH [1]

2 Secondary alcohol [1]

3 There are hydrogen bonds between the water and methanol molecules.

[diagram showing hydrogen bond between methanol and water] [3]

14.2

1 C$_4$H$_9$OH + 6O$_2$ → 4CO$_2$ + 5H$_2$O [1]

2 a i CH$_3$CH$_2$OH + HCl → CH$_3$CH$_2$Cl + H$_2$O

reagents: NaCl, H$_2$SO$_4$ [2]

ii CH$_3$CH$_2$OH + [O] → CH$_3$CHO + H$_2$O

reagents and conditions: K$_2$Cr$_2$O$_7$, H$_2$SO$_4$, distil [2]

iii CH$_3$CH$_2$OH + 2[O] → CH$_3$COOH + H$_2$O

reagents and conditions: K$_2$Cr$_2$O$_7$, H$_2$SO$_4$, reflux [2]

b i substitution, **ii** and **iii** oxidation [2]

3 a Elimination [1]

b C$_4$H$_{10}$O → C$_4$H$_8$ + H$_2$O [1]

c [structures of but-1-ene, E-but-2-ene, Z-but-2-ene] [4]

15.1

1 a The C–X bond polarity decreases down the group as the electronegativity of the halogen decreases from F → I. [2]

b The C–X bond enthalpy decreases down the group. [1]

2 a [nucleophilic substitution mechanism: OH$^-$ attacks CH$_3$Br, displacing Br$^-$] [2]

b Nucleophilic substitution. Heterolytic fission. [2]

Answers to summary questions

3   a   The haloalkanes are warmed with $AgNO_3$(aq) in ethanol. Compare the times for a precipitate of the silver halide to appear. [2]

    b   Iodoalkanes react faster than bromoalkanes, which react faster than chloroalkanes. The rate of hydrolysis increases as the C–X bond enthalpy decreases from C–Cl to C–I. [2]

15.2

1   The ozone layer is in the upper atmosphere. It filters out harmful UV radiation. [1]

2   •Cl radicals form when C–Cl bonds in CFC molecules are broken by homolytic fission in the presence of UV. [1]

    •NO radicals form when nitrogen and oxygen react together at high temperatures and pressures, inside aircraft engines and during lightning strikes in thunderstorms. [1]

3   a   Cl• + O_3 → ClO• + O_2
        ClO• + O → Cl• + O_2 [2]

    b   •NO + O_3 → •NO_2 + O_2
        •NO_2 + O → •NO + O_2 [2]

16.1

1   Reflux is vaporisation followed by condensation back to the original container.

    Distillation is vaporisation followed by condensation into a different container. [2]

2   Redistillation is used to separate two organic liquids that have very similar boiling points. [1]

3   The mixture is added to a separating funnel. The two layers are separated by running off each layer though the tap into different containers. The final traces of water are removed by swirling the organic layer with a drying agent (e.g. anhydrous $MgSO_4$). The dry organic liquid is separated by decanting the liquid from the solid into another flask. [3]

16.2

1   **A**: alkene and alcohol
    B: ketone and carboxylic acid [2]

2   a $CH_3CH=CHCH_3$ + HBr → $CH_3CH_2CHBrCH_3$ [1]

    b $CH_3CH_2CHOHCH_3$ + HBr → $CH_3CH_2CHBrCH_3$ + H_2O
        reagents NaBr/ H_2SO_4 [2]

3   $CH_3CH=CH_2$ + H_2O → $CH_3CHOHCH_3$
    Reagent: H_2O(g)
    Conditions: H_3PO_4/H_2SO_4/acid catalyst
    $CH_3CHOHCH_3$ + [O] → CH_3COCH_3 + H_2O
    Reagents: $K_2Cr_2O_7$/H_2SO_4
    Conditions: reflux [4]

17.1

1   The M+1 peak exists because 1.1% of carbon is present as the carbon-13 isotope. [1]

2   m/z = 43: $C_3H_7^+$; m/z = 57: $C_4H_9^+$ [2]

3   a   Molecular mass = 60;
        Molecular formula = C_3H_8O [2]

    b $CH_3CH_2CH_2OH$ and $CH_3CHOHCH_3$ [2]

    c   The fragment ion will be CH_2OH^+.
        X must be $CH_3CH_2CH_2OH$, because it is the only alcohol with CH_2OH in its molecule. [3]

17.2

1   They could be identified by comparing the IR spectra with a database of known IR spectra to find a match in the fingerprint region. [1]

2   Propan-1-ol would contains a peak at 3200–3600 cm^{-1} from the alcohol O–H bond.

    Propanal would contains a peak at 1630–1820 cm^{-1} from the C=O bond.

    Propanoic acid would contains a peak at 1630–1820 cm^{-1} from the C=O bond and a broad peak at 2500–3300 cm^{-1} for the O–H bond in carboxylic acids. [3]

3   The peak at 1700 cm^{-1} is caused by a C=O bond in aldehydes, ketones, and carboxylic acids. The broad peak between 2500 and 3300 cm^{-1} is caused by an O–H bond in carboxylic acids.

    The substance must be a carboxylic acid.

    Relative mass of C_2H_4O is 12 × 2 + 1 × 4 + 16 = 44. From the molecular ion peak, molecular mass = 88, two empirical formula units. Therefore molecular formula = $C_4H_8O_2$.

    Possible carboxylic acids for $C_4H_8O_2$ = $CH_3CH_2CH_2COOH$ or $(CH_3)_2CHCOOH$. [4]

Answers to practice questions

Chapter 2

1 A [1]   2 D [1]   3 C [1]   4 B [1]

5 a Isotopes are atoms of the same element with different numbers of neutrons and different masses. [1]

  b i Relative isotopic mass is the mass of an isotope relative to one-twelfth of the mass of an atom of carbon-12. [2]

    ii 6.92 [2]

  c i Different isotopes have the same number of electrons [1]

    ii $2Li(s) + 2H_2O(l) \rightarrow 2LiOH(aq) + H_2(g)$
    species and state symbols [1], balance [1]

6 a ^{12}C [1]

  b i ^{69}Ga: 60%, ^{71}Ga: 40% [1]

    ii 69.8 [2]

  c

	protons	neutrons	electrons
^{69}Ga	31	38	31
$^{71}Ga^{3+}$	31	40	28

  [2]

  d i $4Ga(s) + 3O_2(g) \rightarrow 2Ga_2O_3(s)$ [1]
     $2Ga(s) + 3Cl_2(g) \rightarrow 2GaCl_3(s)$ [1]

    ii $2Ga(s) + 3H_2SO_4(aq) \rightarrow Ga_2(SO_4)_3(aq) + 3H_2(g)$
       species and state symbols [1], balance [1]

Chapter 3

1 B [1]   2 C [1]   3 C [1]

4 a $n(HCl) = 0.0512\,mol$ [1]
    $n(SrCO_3) = 0.0256\,mol$ [1]
    mass of $SrCO_3 = 0.0256 \times 147.6 = 3.78\,g$ [1]

  b i Filter [1]

    ii $c = 0.0256 \times \dfrac{1000}{500} = 0.0512\,mol\,dm^{-3}$ [1]

  c $Sr : N : O = \dfrac{48.78}{87.6} : \dfrac{15.59}{14.0} : \dfrac{35.63}{16.0}$
    $= 0.557 : 1.114 : 2.227$ [1]
    Formula = SrN_2O_4 [1]

5 a i $n = \dfrac{pV}{RT}$ [1]
      $n = \dfrac{(99.3 \times 10^3 \times 198 \times 10^{-6})}{(8.314 \times 298)}$ [1]
      $n = 7.94 \times 10^{-3}\,mol$ [1]
      $M = \dfrac{m}{n} = \dfrac{0.222}{7.94 \times 10^{-3}} = 28.0\,g\,mol^{-1}$ [1]

    ii N_2 or CO or HCN or C_2H_4 [1]

  b $n(Ca(NO_3)_2) = \dfrac{0.509}{164.1} = 3.10 \times 10^{-3}\,mol$ [1]
    $n(NO_2 + O_2) = 2.5 \times 3.10 \times 10^{-3} = 7.75 \times 10^{-3}\,mol$ [1]
    volume of gas = $7.75 \times 10^{-3} \times 24000 = 186\,cm^3$ [1]

6 a $CuO(s) + 2HCl(aq) \rightarrow CuCl_2(aq) + H_2O(l)$
    species and state symbols [1], balance [1]

  b $n(CuO) = 0.126\,mol$ [1]
    $n(HCl) = 0.252\,mol$ [1]
    volume of HCl = $\dfrac{.252 \times 1000}{1.75} = 144\,cm^3$ [1]

Chapter 4

1 D [1]   2 C [1]   3 D [1]   4 B [1]

5 a i $n(Ba(OH)_2) = \dfrac{0.1500 \times 30.00}{1000} = 4.50 \times 10^{-3}\,mol$ [1]
      $n(HNO_3) = 2 \times 4.50 \times 10^{-3} = 9.00 \times 10^{-3}\,mol$ [1]
      $c = 9.00 \times 10^{-3} \times 1000/25.00 = 0.360\,mol\,dm^{-3}$ [1]

  b An alkali is a soluble base, releasing OH^- ions. [1]

  c $H^+(aq) + OH^-(aq) \rightarrow H_2O(l)$ [1]

  d $Na_2CO_3(aq) + 2HNO_3(aq) \rightarrow 2NaNO_3(aq) + CO_2(g) + H_2O(l)$ [1]

6 a i $2Al(s) + 6HCl(aq) \rightarrow 2AlCl_3(aq) + 3H_2(g)$ [1]

    ii Al in Al has been oxidised from 0 to +3 in $Al_2(SO_4)_3$ [1]
       H in H_2SO_4 has been reduced from +1 to 0 in H_2 [1]

    iii Bubbles [1] and Al dissolves [1]

  b i $Al_2O_3(s) + 3H_2SO_4(aq) \rightarrow Al_2(SO_4)_3(aq) + 3H_2O(l)$
      species [1] Balance and state symbols [1]

    ii Neutralisation [1]

Chapter 5

1 D [1]   2 C [1]   3 A [1]

4 a

	Number of orbitals	Number of electrons
The 2p sub-shell	3	6
The 3rd shell	9	18

Each row: [1]

  b i $1s^2 2s^2 2p^6 3s^2 3p^6 3d^{10} 4s^2 4p^6$ [1]

    ii $1s^2 2s^2 2p^6 3s^2 3p^6 3d^6$ [1]

  c i Ca_3P_2 [1]

    ii The ions both have the electron configuration $1s^2 2s^2 2p^6 3s^2 3p^6$. [1]
       Isoelectronic means the same number of electrons [1]

5 a i An ionic bond is electrostatic attraction between oppositely charged ions [1]

    ii A covalent bond is a shared pair of electrons that is attracted between the nuclei of the bonded atoms [1]

Answers to practice questions

b i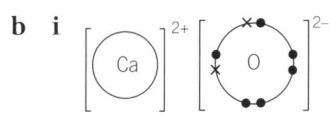

ii 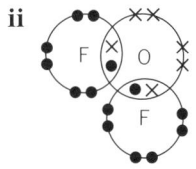 iii 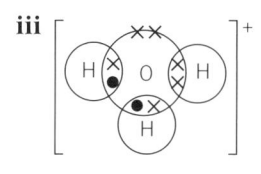 *[4]*

c i The electrostatic attractions between ions are strong *[1]*. High temperatures are needed to provide the energy to break the strong ionic bonds *[1]*.

ii When solid, the ions are fixed in the ionic lattice and cannot move *[1]*. When dissolved, the lattice has broken down and the ions become mobile and are free to move. *[1]*

iii Water is polar *[1]* and the ions are in the lattice are attracted to the dipole charges: Na^+ to $O^{\delta-}$ and Cl^- to $H^{\delta+}$, breaking the ionic bonds. *[1]*

Chapter 6

1 A *[1]*   2 A *[1]*   3 B *[1]*

4 a

Molecule	CO_2	H_2O	NH_3	BH_3
Bonded pairs	4	2	3	3
Lone pairs	0	2	1	0

1 mark for each column *[4]*

b

Molecule	CO_2	H_2O	NH_3	BH_3
Shape	linear	non-linear	pyramidal	trigonal planar
Bond angle	180°	104.5°	107°	120°

1 mark for each column *[4]*

5 a Movement of electrons produces a changing dipole, creating an instantaneous dipole *[1]*

The instantaneous dipole induces a dipole on a neighbouring molecule *[1]*

The induced dipole induces further dipoles on neighbouring molecules, which then attract one another. *[1]*

b i Permanent dipole–dipole interactions *[1]*

ii Cl is more electronegative than H *[1]*

Resulting in a dipole across the molecule: $H^{\delta+}$–$Cl^{\delta-}$ *[1]*. Argon has no dipole. *[1]*

c Hydrogen bonding *[1]*

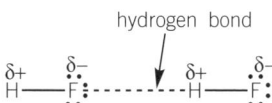

Two HF molecules shown with dipoles *[1]*

Hydrogen bond between lone pair on F of one HF molecule and H on another HF molecule *[1]*

d Order of strength of intermolecular bonds matches order of boiling points:

HF > HCl > Ar *[1]*

Hydrogen bonds are the strongest forces *[1]*

6 a The ability of an atom to attract electrons *[1]* in a covalent bond *[1]*

b $^{\delta+}$O–F$^{\delta-}$ AND $^{\delta-}$O–I$^{\delta+}$ *[1]*

c Most polar: B–F and non-polar: N–Cl *[1]*

d Polar bonds: F is more electronegative than C *[1]*

Non-polar molecule: CF_4 molecule is symmetrical and the dipoles cancel out *[2]*

Chapter 7

1 C *[1]*   2 B *[1]*   3 D *[1]*   4 C *[1]*   5 D *[1]*

6 a i Na, Mg, Al *[1]*   ii Si *[1]*   iii P, S, Cl *[1]*

b Periodicity *[1]*

c Attraction between positive ions and delocalised electrons *[1]*

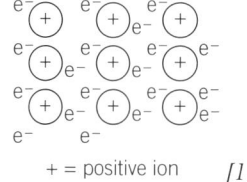

+ = positive ion *[1]*

d For melting weak London forces are broken between P_4 molecules *[1]* In silicon, strong covalent bonds are broken *[1]* Much more energy is required to break stronger forces *[1]*

e Sulfur exists as S_8 molecules and chlorine as Cl_2 molecules *[1]*. London forces are stronger between S_8 molecules as they have more electrons than Cl_2 molecules *[1]*

f Na, Mg and Al conduct AND Si, P, S, Cl do not conduct *[1]*

Conductivity increases from Na to Al *[1]* as number of delocalised electrons increases *[1]*

7 a $C(g) \rightarrow C^+(g) + e^-$ *[1]*

b Across Period 3, nuclear charge increases *[1]*. Electrons are added to the same shell *[1]* and attraction between nucleus and outer electrons increases *[1]*

c In B, electron is removed from a 2p orbital rather than 2s orbital in Be. The 2p sub-shell is at higher energy and its electron is easier to remove *[2]*

d In O, one of the 2p orbitals contains paired electrons whereas in N, all three orbitals are singly occupied. *[1]* The paired electrons in O repel and an electron is easier to remove *[1]*

Answers to practice questions

e i $Na^{6+}(g) \rightarrow Na^{7+}(g) + e^-$ *[1]*

   ii As each electron is removed the same number of protons attract fewer electrons, which are drawn in closer. The remaining electrons are attracted more to the nucleus. *[1]*

   iii There is a large increase between 1st and 2nd ionisation energies and between the 9th and 10th ionisation energies *[1]*, marking electron loss from the next closest shell *[1]*. The element must have 1 electron in outer shell, 8 electrons in a middle shell and 2 electrons in an inner shell. *[1]*

Chapter 8

1 B *[1]*   2 C *[1]*   3 B *[1]*   4 C *[1]*

5 a **A**: calcium oxide *[1]* **B**: calcium hydroxide *[1]*

   $2Ca(s) + O_2(g) \rightarrow 2CaO(s)$ *[1]*

   $CaO(s) + H_2O(l) \rightarrow Ca(OH)_2$ *[1]*

   b 10–14 *[1]*

   c $Ca(OH)_2$ neutralises acid soils *[1]*

   $Ca(OH)_2 + 2HCl \rightarrow CaCl_2 + 2H_2O$ *[1]* (or other acid)

6 a $2Ba(s) + O_2(g) \rightarrow 2BaO(s)$ *[1]*

   $Ba(s) + 2H_2O(l) \rightarrow Ba(OH)_2 + H_2$ *[1]*

   b i $Sr^+(g) \rightarrow Sr^{2+}(g) + e^-$ *[1]*

      ii Down a group, electrons are added to a new shell, further from the nucleus *[1]*

      There are more inner shells between the outer electrons and the nucleus, increasing the shielding *[1]*

      Attraction between nucleus and outer electrons decreases *[1]*

      Outer electrons are lost more easily and ionisation energies decrease. *[1]*

   c i $Mg(s) + 2HCl(aq) \rightarrow MgCl_2(aq) + H_2(g)$

      Species *[1]* Balance and state symbols *[1]*

      ii $n(Ca) = \frac{1.00}{40.1} = 0.0249$ mol *[1]*

      $n(Mg) = \frac{1.00}{24.3} = 0.0412$ mol *[1]*

      volume of H_2 from Ca = 0.0249 × 24000 = 598/599 cm³ *[1]*

      volume of H_2 from Mg = 0.0412 × 24000 = 988/989 cm³ **AND** Student is wrong *[1]*

Chapter 9

1 B *[1]*   2 C *[1]*

3 a $CS_2(l) + 3O_2(g) \rightarrow CO_2(g) + 2SO_2(g)$ *[2]*

   b $q = mc\Delta T = 150 \times 4.18 \times 11.5$

   = 7210.5 J = 7.2105 kJ *[1]*

   $n(CS_2)$ burnt = $\frac{0.9525}{76.2}$ = 0.0125 mol *[1]*

   For 1 mol H_2O, energy change = $\frac{7.2105}{0.0125}$ = 576.84 kJ.

   ∴ $\Delta_c H = -577$ kJ mol⁻¹

   1 mark for value, 1 mark for sign *[2]*

   c Heat loss *[1]*; Incomplete combustion *[1]*

4 a i Standard enthalpy change of formation is the enthalpy change for the formation of 1 mole *[1]* of a substance from its constituent elements *[1]* under standard conditions of 100 kPa and 298 K *[1]*

      ii Formation of $O_2(g)$ from $O_2(g)$ is no change *[1]*

      iii $\Delta_r H = [5 \times -393.5 + 6 \times -285.8] - [-173.2 + 0]$

      = -3509.1 kJ mol⁻¹

      Use of 5 and 6 *[1]*; Correct subtraction *[1]*

      Correct answer *[1]*

   b i The enthalpy change for breaking a particular bond *[1]* in one mole of bonds in gaseous molecules *[1]*

      ii Σ(Bond enthalpies of reactants) = N–N + 4(N–H) + 2(O–O) + 4(O–H) = +158 + (4 × +391) + (2 × +144) + (4 × +464)

      = 3866 kJ *[1]*

      $\Delta_r H$ = Σ(Bond enthalpies of reactants) – Σ(Bond enthalpies of products)

      ∴ $-787 = 3866 - [N≡N + 8 \times O–H]$

      ∴ $-787 = 3866 - [N≡N + 8 \times +464]$ *[1]*

      ∴ bond enthalpy N≡N = 3866 − 3712 + 787

      = +941 kJ mol⁻¹ *[1]*

Chapter 10

1 C *[1]*   2 C *[1]*   3 D *[1]*

4 a Concentration is increased *[1]*, there are more frequent collisions and the rate increases *[1]*

   b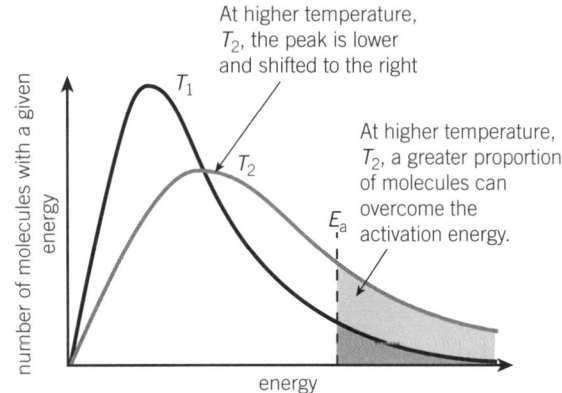

   Correct axes *[1]*

   Boltzmann distribution at **two** different temperatures starting at the origin AND not touching x axis at high energy *[1]*

   Higher temperature curve (T_2) has lower maximum AND maximum shifted to higher energy *[1]*

   At the higher temperature more molecules have energy above activation energy *[1]*

Answers to practice questions

c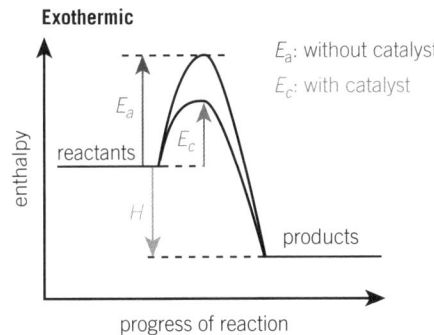

Products below reactant **AND** ΔH labelled with product **AND** arrow downwards [1]

E_a labelled correctly, above reactants [1]

E_c labelled correctly, and $< E_a$ [1]

5 a Rate of forward reaction = rate of reverse reaction [1] Concentrations are constant [1]

b Equilibrium position shifts to left [1] as the formation of products is exothermic and gives out energy [1]

c Equilibrium position shifts to right [1] as there are fewer gaseous molecules on the right [1]

d i $K_c = \dfrac{[NO_2(g)]^2}{[NO(g)]^2[O_2(g)]}$ [1]

ii $[NO_2(g)] = \sqrt{K_c \times [NO(g)]^2 \times [O_2(g)]}$

$= \sqrt{1.57 \times 10^5 \times (3.0 \times 10^{-3})^2 \times 1.5 \times 10^{-3}}$

$= 4.6 \times 10^{-2}$ mol dm^{-3} [2]

Chapter 11

1 C [1]   2 B [1]   3 B [1]   4 D [1]

5 a i Compounds with no multiple carbon–carbon bonds [1] that contain carbon and hydrogen only [1]

ii A, B, E, and F [1]

b A homologous series contains organic compounds with the same functional group [1] but with each successive member differing by CH$_2$ [1]

A functional group is a group of atoms responsible for the characteristic reactions of a compound [1]

c i A, B, C, E, and F [1]   ii E and F [1]

d D [1]

e i Compounds with the same molecular formula but different structural formulae [1]

ii C, E, and F [1]

f i Addition [1] alcohol [1]

ii Substitution [1] bromoalkane OR haloalkane [1]

iii Elimination [1] alkene [1]

6 a C : H = $\dfrac{85.71}{12.0} : \dfrac{14.29}{1.0}$ = 7.14 : 14.29 [1] Empirical formula = CH$_2$ [1]

Molecular formula = CH$_2 \times \dfrac{56}{14}$ = C$_4$H$_8$ [1]

b i

ii Left to right: but-1-ene, but-2-ene, methylpropene, methylcyclopropane [1] for each structure with name

Chapter 12

1 B [1]   2 D [1]   3 C [1]   4 D [1]

5 a CH$_3$CH$_2$CH$_3$ → CH$_3$CH$_2$CH$_2$Br + HBr [1]

b i Radical substitution [1]

ii A covalent bond breaks with each bonded atom taking one of the shared pair of electrons from the bond to form two radicals. [1]

iii Initiation   Br$_2$ → 2Br• [1]

Propagation   C$_3$H$_8$ + Br• → C$_3$H$_7$• + HBr [1]

C$_3$H$_7$• + Br$_2$ → C$_3$H$_7$Br + Br• [1]

Termination   2C$_3$H$_7$• → C$_6$H$_{14}$

2Br• → Br$_2$

C$_3$H$_7$• + Br• → C$_3$H$_7$Br

All 3 terminations [2]; 2 terminations [1]

Initiation, propagation and termination linked to correct equations [1]

c Further substitution [1], e.g. to form CH$_3$CH$_2$CHBr$_2$ [1]

Substitutions at different positions on the carbon chain [1], e.g. CH$_3$CHBrCH$_3$ [1]

6 a C$_{32}$H$_{66}$ [1]   b 2,2,4-trimethylpentane [1]

c From propane to butane to pentane, the carbon chain length increases [1] and there will be a more surface contact is between molecules AND the London forces between the molecules will be greater [1] and so more energy is required to overcome the intermolecular forces, increasing the boiling point [1]

From pentane to 2-methylbutane to 2,2-dimethylpropane, branching increases [1] Less surface contact AND weaker London forces [1], decreasing the energy needed to break the intermolecular forces. [1]

d i C$_8$H$_{18}$(l) + 12½ O$_2$(g) → 8CO$_2$(g) + 9H$_2$O(l) species [1] balance + state symbols: (l) or (g) for H$_2$O [1]

ii $n(CO_2) = \dfrac{72.0}{24.0} = 3.00$ mol [1]

$n(C_8H_{18}) = \dfrac{3}{8} = 0.375$ mol [1]

mass C$_8$H$_{18}$ = 0.375 × 114 = 42.75 g [1]

$n(O_2)$ = 0.375 × 12.5 = 4.6875 mol [1]

volume of air = 4.6875 × 24.0 × 5 = 562.5 dm^3 [1]

Answers to practice questions

Chapter 13

1 B [1]   2 D [1]   3 A [1]   4 A [1]

5 a

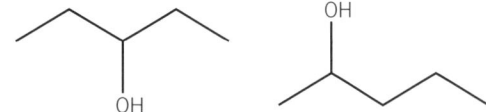

A [1]   B [1]   C [1]   D [1]

b Nickel [1]

c Curly arrow from C=C to H$^{\delta+}$ of HBr [1]

Curly arrow from H–Br and correct dipole [1]

Correct carbocation AND curly arrow from Br⁻ to C⁺ [1]

The primary carbocation is less stable [1]

(CH$_3$)$_3$CBr is major product [1]

d i

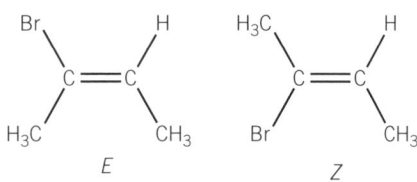

ii Addition [1]

iii Use as a feedstock for plastics and organic chemicals [1]

Combustion for energy production [1]

6 a i C : H = $\frac{85.71}{12.0} : \frac{14.29}{1.0}$ = 7.14 : 14.29 [1]

ii Empirical formula = CH$_2$ [1]

Molecular formula = CH$_2$ × $\frac{70}{14}$ = C$_5$H$_{10}$ [1]

iii

E isomer   [1]

Pent-2-ene is the only isomer of C$_5$H$_{10}$ that has stereoisomers [1]. *E* isomer has higher priority groups on opposite sides of the C=C bond [1]

b i Nickel [1]

ii Structure of **G**

[1]

c i Acid catalyst (phosphoric or sulfuric acid) [1]

ii Structure of **H** and **I** [1] for each structure

d

E   *Z*

Correct structures [1] Correct labels [1]

Highest priority groups are Br on left and CH$_3$ on right as Br and C have larger atomic numbers. [1]

Chapter 14

1 B [1]   2 A [1]   3 D [1]

4 Reagents: Acid/H⁺ and dichromate/Cr$_2$O$_7^{2-}$ [1]

Observations: Orange to Green/blue [1]

Distillation produces aldehyde [1]

CH$_3$CH$_2$CH$_2$CH$_2$OH + [O] → CH$_3$CH$_2$CH$_2$CHO + H$_2$O

organic product [1] complete equation [1]

Reflux produces carboxylic acid [1]

CH$_3$CH$_2$CH$_2$CH$_2$OH + 2[O] → CH$_3$CH$_2$CH$_2$COOH + H$_2$O

organic product [1] complete equation [1]

5 a 2,3-dimethylbutan-1-ol [1]   b C$_6$H$_{14}$O [1]

c

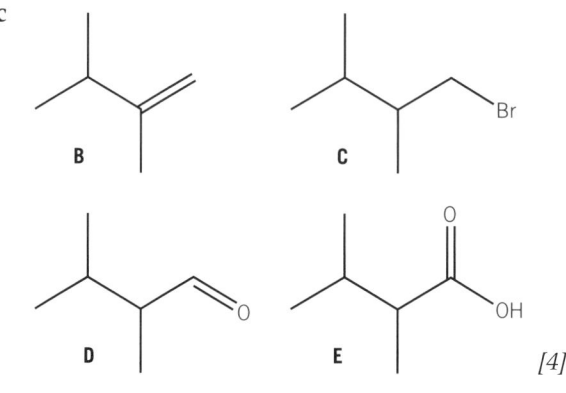

[4]

d B: elimination [1]; C: substitution [1]

D and E: oxidation [1]

e H$_2$O [1]   f F

[1]

6 a NaBr/H$_2$SO$_4$ [1]

b n(RBr) = $\frac{10.0}{164.9}$ = 0.06064 mol [1]

n(ROH) required = $\frac{100}{40}$ × 0.06064 = 0.1516 mol [1]

mass ROH needed = 0.1516 × 102 = 15.5 g [1]

Chapter 15

1 D [1]   2 C [1]   3 A [1]   4 A [1]

5 a CH$_3$CH$_2$CH$_2$CH$_2$Br + OH⁻ → CH$_3$CH$_2$CH$_2$CH$_2$OH + Br⁻ [1]

b A nucleophile donates an electron pair [1]

c Dipole shown on the C–Br bond and curly arrow from the C–Br bond to the Br atom [1]

Curly arrow from lone pair or negative charge on :OH⁻ to carbon atom in the C–Br bond [1]

Correct organic product and Br⁻ [1]

d Heterolytic fission [1]

Answers to practice questions

e Rate increases from 1-chlorobutane to 1-bromobutane to 1-iodobutane [1]

Bond enthalpy decreases C–Cl > C–Br > C–I [1]

f i Correct structure [1]

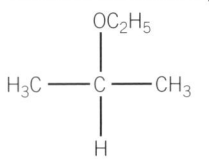

ii Correct structure [1]

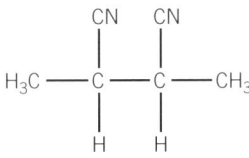

6 a Ozone prevents harmful UV from reaching Earth which would cause skin cancers and damage life [1]

b i Correct structure [1]

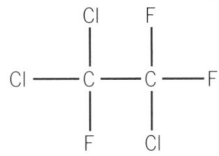

ii UV radiation breaks C–Cl bonds to form Cl• radicals [1]

$C_2Cl_3F_3 \rightarrow C_2Cl_2F_3\bullet + Cl\bullet$ [1]

iii $Cl\bullet + O_3 \rightarrow ClO\bullet + O_2$ [1]

$ClO\bullet + O \rightarrow Cl\bullet + O_2$ [1]

overall $O_3 + O \rightarrow 2O_2$ [1]

Chapter 16

1 a $CH_3(CH_2)_3OH + HBr \rightarrow CH_3(CH_2)_3Br + H_2O$ [1]

b i $n(CH_3(CH_2)_3Br) = \frac{5.00}{136.9} = 0.03652$ mol [1]

$n(CH_3(CH_2)_3OH)$ required $= \frac{100}{55} \times 0.03652$
$= 0.06641$ mol [1]

mass ROH needed $= 0.06641 \times 74.0 = 4.91$ g [1]

ii $n(NaBr) = 0.06641 \times 102.9 = 6.83$ g [1]

$n(H_2SO_4) = 0.06641 \times 98.1 = 6.51$ g [1]

c i Vaporisation followed by condensation back to the same container [1]

ii Connect a condenser vertically above a pear-shaped or round-bottom flask [1]

iii Enables a slow reaction to be left boiling for a length of time [1] while preventing the contents of flask from boiling dry [1]

d Add the mixture to a separating funnel [1]

As the 1-bromobutane has a greater density than water, run out the lower layer [1]

e Add anhydrous $MgSO_4$, which removed traces of water [1] Decant the organic liquid from the solid/hydrated salt [1]

f Distil [1] and collect the fraction that boils at 101–103°C, the boiling point of 1-bromobutane [1]

2 a i **A**: bromoalkane [1]; **B**: aldehyde [1]

ii **1st stage**: reactants NaOH(aq) [1]

[structure: propyl bromide + OH⁻ → propanol + Br⁻, labelled A]

Organic product [1], full equation [1]

2nd stage: reactants $K_2Cr_2O_7/H_2SO_4$, distil [1]

[structure: propanol + [O] → propanal + H_2O, labelled B]

Equation [1]

b i **C**: alkene and alcohol [1]

D: ketone and carboxylic acid [1]

ii **1st stage**: reactants $H_2O(g)$ and acid catalyst [1]

[structure: allyl alcohol + H_2O → propane-1,2-diol, labelled C]

Organic product [1], full equation [1]

2nd stage: reactants $K_2Cr_2O_7/H_2SO_4$, reflux [1]

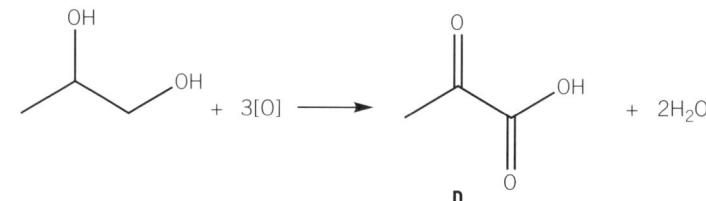

Equation [1]

Chapter 17

1 D [1]   2 D [1]   3 A [1]

4 $C : H : O = \frac{60.00}{12.0} : \frac{13.33}{1.0} : \frac{26.67}{16.0}$ [1]

$= 5.00 : 13.33 : 1.67 = C_3H_8O$ [1]

From mass spectrum, $M_r = 60$ which matches C_3H_8O [1]

Peak at 1720 cm⁻¹ indicates presence of C=O [1]

Peak at 2500–3300 cm⁻¹ indicates the presence of –OH group in COOH [1]

Structure of **A** and **B** [1 mark for each = 2]

[structures: A = propan-1-ol (CH₃CH₂CH₂OH); B = propanoic acid (CH₃CH₂COOH)]

$CH_3CH_2CH_2OH + 2[O] \rightarrow CH_3CH_2COOH + H_2O$ [1]

Peak **X**: CH_2OH^+ [1]

Periodic table

(1)	(2)											(3)	(4)	(5)	(6)	(7)	(0)
1																	**18**
1 **H** hydrogen 1.0																	2 **He** helium 4.0
												13	14	15	16	17	
3 **Li** lithium 6.9	4 **Be** beryllium 9.0											5 **B** boron 10.8	6 **C** carbon 12.0	7 **N** nitrogen 14.0	8 **O** oxygen 16.0	9 **F** fluorine 19.0	10 **Ne** neon 20.2
11 **Na** sodium 23.0	12 **Mg** magnesium 24.3	3	4	5	6	7	8	9	10	11	12	13 **Al** aluminium 27.0	14 **Si** silicon 28.1	15 **P** phosphorus 31.0	16 **S** sulfur 32.1	17 **Cl** chlorine 35.5	18 **Ar** argon 39.9
19 **K** potassium 39.1	20 **Ca** calcium 40.1	21 **Sc** scandium 45.0	22 **Ti** titanium 47.9	23 **V** vanadium 50.9	24 **Cr** chromium 52.0	25 **Mn** manganese 54.9	26 **Fe** iron 55.8	27 **Co** cobalt 58.9	28 **Ni** nickel 58.7	29 **Cu** copper 63.5	30 **Zn** zinc 65.4	31 **Ga** gallium 69.7	32 **Ge** germanium 72.6	33 **As** arsenic 74.9	34 **Se** selenium 79.0	35 **Br** bromine 79.9	36 **Kr** krypton 83.8
37 **Rb** rubidium 85.5	38 **Sr** strontium 87.6	39 **Y** yttrium 88.9	40 **Zr** zirconium 91.2	41 **Nb** niobium 92.9	42 **Mo** molybdenum 95.9	43 **Tc** technetium	44 **Ru** ruthenium 101.1	45 **Rh** rhodium 102.9	46 **Pd** palladium 106.4	47 **Ag** silver 107.9	48 **Cd** cadmium 112.4	49 **In** indium 114.8	50 **Sn** tin 118.7	51 **Sb** antimony 121.8	52 **Te** tellurium 127.6	53 **I** iodine 126.9	54 **Xe** xenon 131.3
55 **Cs** caesium 132.9	56 **Ba** barium 137.3	57–71 lanthanoids	72 **Hf** hafnium 178.5	73 **Ta** tantalum 180.9	74 **W** tungsten 183.8	75 **Re** rhenium 186.2	76 **Os** osmium 190.2	77 **Ir** iridium 192.2	78 **Pt** platinum 195.1	79 **Au** gold 197.0	80 **Hg** mercury 200.6	81 **Tl** thallium 204.4	82 **Pb** lead 207.2	83 **Bi** bismuth 209.0	84 **Po** polonium	85 **At** astatine	86 **Rn** radon
87 **Fr** francium	88 **Ra** radium	89–103 actinoids	104 **Rf** rutherfordium	105 **Db** dubnium	106 **Sg** seaborgium	107 **Bh** bohrium	108 **Hs** hassium	109 **Mt** meitnerium	110 **Ds** darmstadtium	111 **Rg** roentgenium	112 **Cn** copernicium		114 **Fl** flerovium		116 **Lv** livermorium		

Key
atomic number
Symbol
name
relative atomic mass

57 **La** lanthanum 138.9	58 **Ce** cerium 140.1	59 **Pr** praseodymium 140.9	60 **Nd** neodymium 144.2	61 **Pm** promethium 144.9	62 **Sm** samarium 150.4	63 **Eu** europium 152.0	64 **Gd** gadolinium 157.2	65 **Tb** terbium 158.9	66 **Dy** dysprosium 162.5	67 **Ho** holmium 164.9	68 **Er** erbium 167.3	69 **Tm** thulium 168.9	70 **Yb** ytterbium 173.0	71 **Lu** lutetium 175.0
89 **Ac** actinium	90 **Th** thorium 232.0	91 **Pa** protactinium	92 **U** uranium 238.1	93 **Np** neptunium	94 **Pu** plutonium	95 **Am** americium	96 **Cm** curium	97 **Bk** berkelium	98 **Cf** californium	99 **Es** einsteinium	100 **Fm** fermium	101 **Md** mendelevium	102 **No** nobelium	103 **Lr** lawrencium

Data sheet

General Information

Molar gas volume = 24.0 dm³ mol⁻¹ at room temperature and pressure, RTP

Avogadro constant, $N_A = 6.02 \times 10^{23}$ mol⁻¹

Specific heat capacity of water, $c = 4.18$ J g⁻¹ K⁻¹

Ionic product of water, $K_w = 1.00 \times 10^{-14}$ mol² dm⁻⁶ at 298 K

1 tonne = 10^6 g

Arrhenius equation: $k = Ae^{-E_a/RT}$ or $\ln k = -E_a/RT + \ln A$

Gas constant, $R = 8.314$ J mol⁻¹ K⁻¹

Characteristic infrared absorptions in organic molecules

Bond	Location	Wavenumber / cm⁻¹
C–C	Alkanes, alkyl chains	750–1100
C–X	Haloalkanes (X = Cl, Br, I)	500–800
C–F	Fluoroalkanes	1000–1350
C–O	Alcohols, esters, carboxylic acids	1000–1300
C=C	Alkenes	1620–1680
C=O	Aldehydes, ketones, carboxylic acids, esters, amides, acyl chlorides and acid anhydrides	1630–1820
aromatic C=C	Arenes	Several peaks in range 1450–1650 (variable)
C≡N	Nitriles	2220–2260
C–H	Alkyl groups, alkenes, arenes	2850–3100
O–H	Carboxylic acids	2500–3300 (broad)
N–H	Amines, amides	3300–3500
O–H	Alcohols, phenols	3200–3600

Data sheet

^{13}C NMR chemical shifts relative to TMS

^{1}H NMR chemical shifts relative to TMS

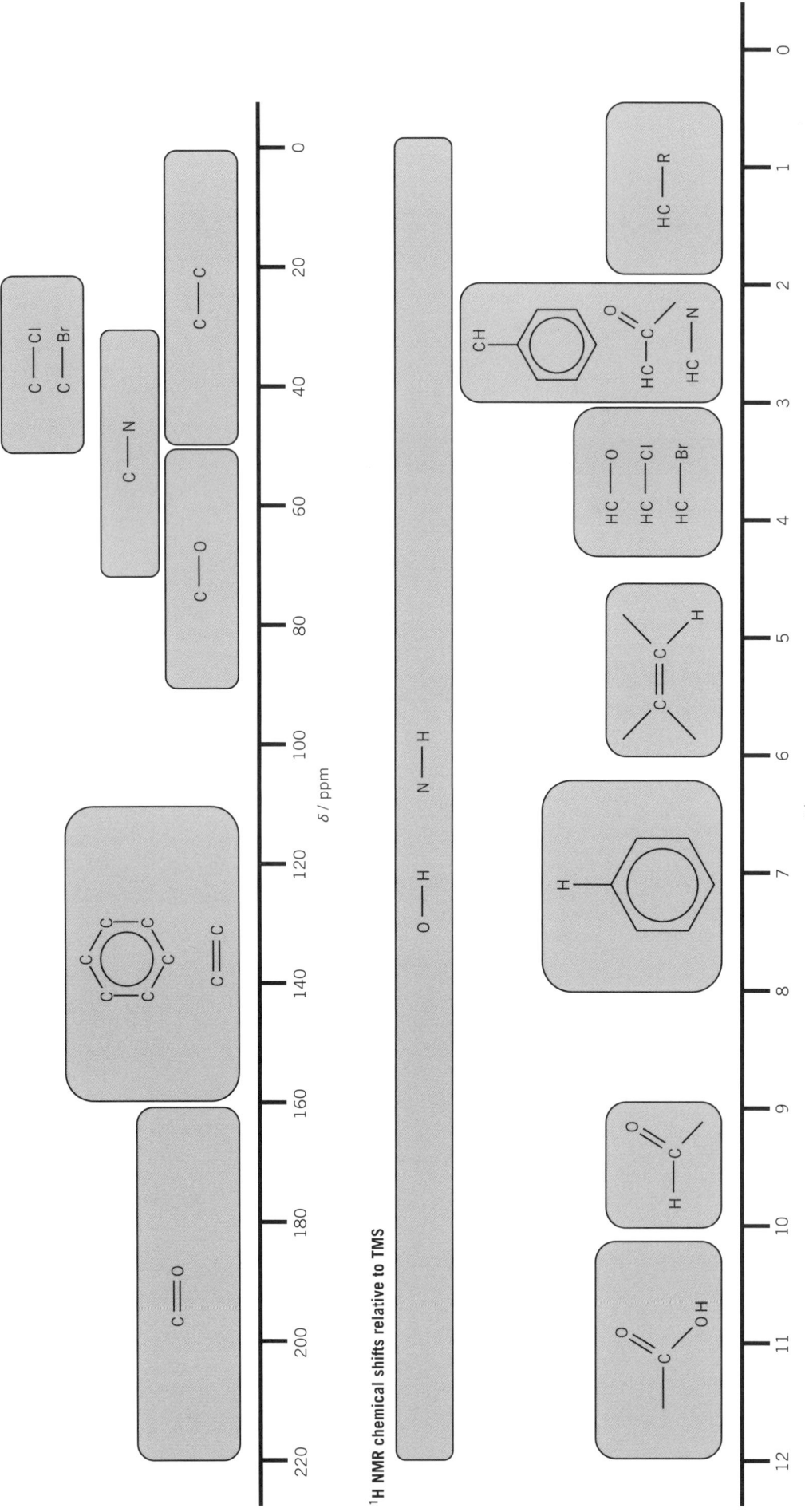

Chemical shifts are variable and can vary depending on the solvent, concentration and substituents. As a result, shifts may be outside the ranges indicated above.

OH and **NH** chemical shifts are very variable and are often broad. Signals are not usually seen as split peaks.

Note that **CH** bonded to 'shifting groups' on either side, e.g. $O-CH_2-C=O$, may be shifted more than indicated above.